SEED PROCESSING WORKER

SEED PROCESSING WORKER

by
Chithra R Nair
Deepak Gopalakrishnan

Wordpen Academics
Publishing Science That Matter

ISBN: 978-93-6084-247-5

Published by : **WORDPEN ACADEMICS**
4736/23, Ansari Road, Darya Ganj,
New Delhi – 110 002
Phone: +91-011-35004336
E-mail: info@wordpenacademics.com

Printed at **:** **Replika Press Pvt. Ltd.**

Exclusive Distributor: ***Bio-Green Books, New Delhi***

Preface

Seed Processing Worker is a comprehensive guide designed to provide essential knowledge and practical skills to individuals involved in the seed processing industry. Seed processing is a vital part of agriculture, as it ensures the production of high-quality seeds that are free from diseases, impurities, and genetic inconsistencies. The work of seed processing workers directly influences crop yields, food security, and the sustainability of agricultural practices worldwide.

This book is aimed at students, workers, and professionals in the seed processing and agricultural sectors, as well as those interested in understanding the importance of seed quality management in modern farming systems. Whether you are beginning your career in seed processing or are seeking to deepen your understanding of the industry, this text will serve as both an instructional resource and a reference guide.

The book begins with an overview of the role of seed processing in agriculture, discussing the different stages of seed production, from harvesting to storage. It then explores the key activities in seed processing, such as cleaning, grading, treating, and packaging. Throughout the chapters, readers will learn about the equipment and technologies used in seed processing, including seed cleaners, air separators, graders, and packaging machines.

A significant portion of the book focuses on the quality control processes in seed processing. Ensuring that seeds meet required standards for germination, purity, and vigor is crucial for successful crop production. Topics such as seed testing, moisture content analysis, and the prevention of seed-borne diseases are discussed in detail, with practical tips on how to identify and mitigate potential quality issues.

The book also highlights the importance of seed certification and the role of regulatory bodies in ensuring that processed seeds meet national and international standards. It explores the ethical and environmental considerations related to seed production, such as sustainable farming practices and the impact of genetic modification in seed development.

Real-world examples, case studies, and exercises are provided to help readers apply their knowledge in practical settings. The text also includes a section on career opportunities in the seed industry, providing insight into how seed processing workers can advance in their professional journeys.

We hope that *Seed Processing Worker* will not only provide readers with technical expertise but also inspire them to appreciate the crucial role of seed quality in ensuring global food security and agricultural sustainability.

Chithra R Nair

Deepak Gopalakrishnan

Contents

Unit 1: Clean and Dry Seeds

1

Receive Seeds from Growers

1.1 Receive seeds from various seed growers as per the SOP

1.1.1 Seeds

Seed is the basic and most critical input for sustainable agriculture. Quality seeds are the most important farming inputs to achieve high yields and good produce can be achieved only when the seeds are used wisely. For seed processing worker, access to affordable and high quality seeds is therefore important for increasing agricultural production and productivity. In many developing countries, the formal seed system has shifted from a publicly dominated sector to a privately dominated sector.

Quality seeds is another input necessary for agricultural productivity, and good quality seeds account for 20%-25% of increased crop productivity.

1.1.2 Types of Seed

A Seed is primarily of two types:

- Monocotyledonous Seed
- Dicotyledonous Seed

Monocotyledonous Seed

A Monocotyledonous seed, as the name suggests, has only one cotyledon. There is only one outer layering of the seed coat. A seed has the following parts:

- Seed Coat: In the seed of cereals such as maize, the seed coat is membranous and generally fused with the fruit wall, called Hull.
- Endosperm: The endosperm is bulky and stores food. Generally, monocotyledonous seeds are endospermic but some as in orchids are non-endospermic.
- Aleuron layer: The outer covering of endosperm separates the embryo by a proteinous layer called aleurone layer.
- Embryo: The embryo is small and situated in a groove at one end of the endosperm.
- Scutellum: This is one large and shield-shaped cotyledon.
- Embryonal axis: Plumule and radicle are the two ends.
- Coleoptile and coleorhiza: The plumule and radicle are enclosed in sheaths. They are coleoptile and coleorhiza.

Dicotyledonous Seed

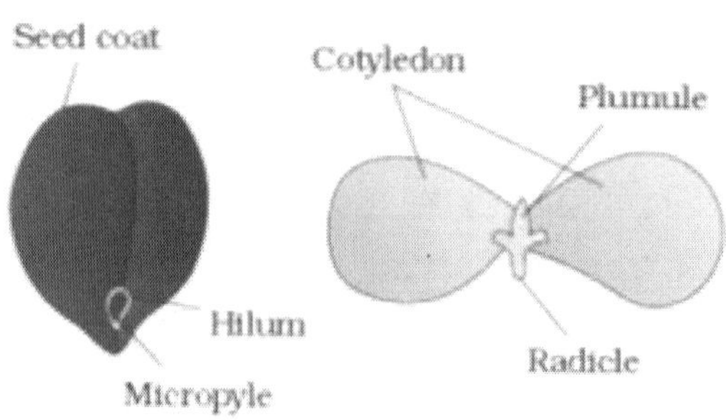

Unlike monocotyledonous seed, a dicotyledonous seed, as the name suggests, has two cotyledons. It has the following parts:

- Seed coat: This is the outermost covering of a seed. The seed coat has two layers, the outer testa and the inner tegmen.
- Hilum: The hilum is a scar on the seed coat through which the developing seed was attached to the fruit.
- Micropyle: It is a small pore present above the hilum.
- Embryo: It consists of an embryonal axis and two cotyledons.
- Cotyledons: These are often fleshy and full of reserve food materials.
- Radicle and plumule: They are present at the two ends of the embryonal axis.

- Endosperm: In some seeds such as castor, the endosperm formed as a result of double fertilisation, is a food storing tissue. In plants such as bean, gram and pea, the endosperm is not present in the matured seed. They are known as non-endospermous.

1.1.3 Seed Processing Worker

Seed processing worker is a person who is responsible in doing following functions such as:

- knowledge about different types of seeds and their characteristics
- knowledge related to amount of moisture content required for good storage life of seeds
- knowledge related to different methods of cleaning and drying of seeds
- knowledge about drying and treating of seeds after cleaning them properly
- taking sample seeds for testing
- storing of different seeds in an appropriate manner
- packing of different seeds for distribution
- assuring about the quality of seed processing

1.1.4 Seed Grower

Seed Growers are artisanal farmers that worked together for many years to grow, develop, and bring high quality, locally grown and adapted seeds.

For growing seeds, seed growers should:

- get required seed information and details about growing seed rules/policies from necessary authorities.
- read and follow different growing strategies by discussing with concerned authorities.
- able to know the correct seed information for growing particular seed crop.
- understand seed details in terms of purity, usage, growing strategies.
- get information from state's department of agriculture/certifying agencies on where to buy registered seeds.

1.1.5 Seed producing organizations

There are three types of Seed Producing Agencies such as:

- Government sector
- Public sector
- Private sector

Public sector organizations include:

National Seeds Corporation (NSC)

NSC was established in 1963 which involves in work of quality seed production. Its responsibilities are defined with the inception of National Seeds Project (NSP) and its main function involves in:

- Production of Breeder Seed: The requirement of breeder seed of various crops is assessed and arrangements are made for its production through crop research institutes of ICAR and State Agricultural Universities.
- Production of Foundation Seed: The production and distribution of the foundation seed of the varieties of national importance is the responsibility of the NSC.
- Production of Certified Seed: The production and marketing of the certified seeds of vegetables, oilseeds, pulses and fodder crops is taken up.
- Marketing: Interstate marketing of the seeds of national varieties lies with the NSC.
- Consultancy Services: The NSC provides consultancy services related to designing, procurement, and installation of seed processing plants, equipments, etc.
- Imparts Training: The NSC organizes short-term training courses to the personnel engaged in seed production work.
- Coordination: NSC provides coordination in the production of certified seed by SSCs.
- Taking up seed certification work in those states where SSCA has not been established.
- Import and export work of seed.

State Seed Corporations (SSCs)

National Seed Project in 1976 establishes State Seed Corporations in 14 states of India. Its functions involve in production and distribution of certified seed in respective state. With the establishment of Tarai Development Corporation in U.P. in 1969 for production of certified seeds, SSCs has been established in other states based on similar pattern to faster the development of seed industry in the country. NSP provides finance for establishment of new SSCs and strengthening of the existing ones. The NSP provides funds for following purposes:

- Strengthening seed production, processing and marketing facilities of NSC.
- Strengthening the facilities of the NSC for producing and processing seeds of vegetables, fodder crops and other specialised crops not taken up by the SSCs.

- Strengthening the facilities for nucleus, breeder and foundation seed production; processing, and storage at selected Central Research Institute of ICAR and at SAUs.

State Seed Certification Agencies (SSCAs): State Seed Certification Agencies are responsible for seed certification in states. SSCAs performs following functions:

- screen applications from seed growers for seed certification and decides on their fitness,
- check and verify appropriateness of source seed used for growing seed crop under certification,
- do requisite field inspections,
- conduct seed tests,
- certify seeds found suitable and issue appropriate tags for certified and foundation seeds,
- guide seed growers on production, processing and distribution of seeds,
- conduct short courses on seed production for seed growers,
- participate in activities conducted to development of seed industry,

Private sector includes various seed companies.

There are many private seed companies which involves in work of quality seed production such as:

- Maharashtra Hybrid Corporation, Jalna, 1964.
- Pioneer Seed Company, Hyderabad.
- Nath Seed Company Pvt. Ltd., Aurangabad.
- Hindustan Lever Ltd., Mumbai.
- Proagro Seed Company, New Delhi.
- Ankur Seed Company, Nagpur.
- Vijay Seed Company, Jalna.
- Mahendra Seed Company, Jalna.

1.1.6 Receive/access seeds from Seed Growers

Seed processing worker receives different seeds from artisanal farmers/ organization for processing. Before keeping the respective seeds in seed bin, seed processing worker will do the inspection of seed bins which covers:

- Disinfecting of inspected equipment (boots, knife, etc.).
- Observing general conditions in storage
- Temperature
- Airflow

- Humidity
- Presence of moisture or frost in wall or ceiling
- Sanitation practices

Empty received seeds in labelled bin

Seeds after receiving from different seed growers are kept in respective seed bins. Once the inspection of seed bins is done, seed processing worker should keep different seeds in different bins as per the label marked on bins.

While collecting seeds in seed bins, seed processing worker should keep the note of following information:

- Grower's name
- Seed variety name
- Seed type
- Maturity status/anticipated number of growing days
- Seed status-certified/normal
- Class (assigned/already assigned);
- Number of field sample units (number of stems or tubers);
- Date planted (for stem samples only);
- Date collected
- Grower's detail

Assessing seeds by visually inspection

Seeds collected in respective seed bin needs to be visually inspected by seed inspector/processing worker. Seed inspections starts with overall examination of consignment, container, packaging and means of conveyance to get indications about:

- adverse conditions like temperature, moisture content
- physical condition of seed
- abnormality in seeds (hard, damaged, deformed or decayed seeds)
- live or dead insects or contamination

1.2 Checking for immature, shrivelled, broken, diseased, moisture-damaged, infested, insect-damaged seed of seed crop, seed of weeds and wild plants etc.

- Seed processing worker during seed inspection will check for immature, shrivelled, broken, diseased, moisture-damaged, infested, insect-damaged seeds.
- Physical quality parameters such as seed uniformity, extent of inert material content and discoloured seed can be detected by visually examining seed samples.
- Closely examining handfuls of seed is the first step for a better understanding of the quality of seed provided to farmers; it gives the first but not the only opportunity to decide on seed cleaning needs.
- Physical qualities of seed in seed lot can be characterized by inspection of minimal:
  - damaged seed: Damaged seeds includes broken, cracked or shrivelled seed needs to be checked as such seeds will not germinate and is more likely to be attacked by insects or micro-organisms, so such seeds should be eliminated.
  - amount of weed seed or inert matter: Good quality seed should be free of weed seeds, chaff, stones, dirt and seed of other crops, so all such impurities should be removed before processing/conditioning.
  - diseased seed: Discoloured or stained seed carry micro-organisms which can attacks the seed when it starts grow, as plant may live and spread such disease to other plants.
  - uniform seed size: Mature medium and large seed have high germination and vigour as compared to small and immature seed, so while conditioning of seed lot, undersized and light seeds should be eliminated.

If the seeds are found immature, shrivelled, broken, diseased, moisture-damaged, infested, insect-damaged, in such case seed processing worker should immediately return such seeds to respective seed growers.

1.3 Take samples from seeds received for testing if they look visibly acceptable

1.3.1 Seed Examination

- Physical examination will help seed processing worker in analysing and confirming that seeds are free of contamination and from contaminants like weed seeds, other variety of seeds, inert material like plastics, papers, metal etc, damaged seeds and completely off sized seeds.
- Observing physical characteristics of seeds is an important step of checking and determining whether a seed lot is good or not.
- Unlike physical examination of seeds except for mould, it is not always very easy for farmers/growers to spot these when they examine seeds as most of them are invisible to the naked eyes.
- Apart from observing moulds some of which are colourful and feeding on outer coat of seeds, many disease-carrying seeds cannot be easily detected.
- Seed processing workers/crop growers can use physical senses of seeing, touching and feeling, so as to determine the quality of seed by considering following:
  - size of seeds
  - length of seed
  - weight of seed
  - shape of seed
  - surface texture of seed
  - colour of seed

1.3.2 Taking Seed Sample

Taking seed sample manually

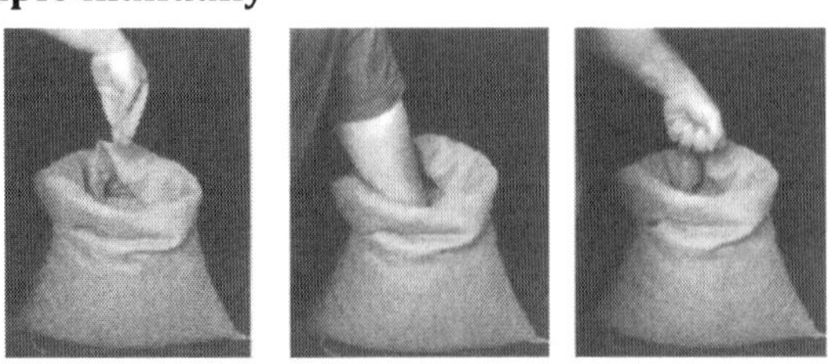

Seed processing worker should follow basic steps in taking seed sample from seed lot by:

- ❖ push the hand into the seed container
- ❖ closing the hand with seeds
- ❖ withdraw the hand by taking great care so that the fingers remain tightly closed around the seeds so that seeds may not fall
- ❖ use gloves for taking out treated seeds

Once the seeds are physically examined by seed processing worker kept in the respective seed bins, the seeds are taken out from the bags for testing. The seed samples are arranged in respective testing boxes.

To have sample of seed size suitable for tests, in which the probability of constituent being present can be determined by its level of occurrence in seed lot.

1.3.3 Seeds for Testing

- ❖ Seed testing is done for research purposes or to determine, if seed storage techniques are functioning.
- ❖ Testing of seeds is done to find the planting value of seeds and in determining:
  - seed quality suitable for planting
  - seed quality problems and its cause
  - need for drying and processing and specific procedures to be used
  - seed quality standards or labelling specifications
  - seed quality and provides basis for price and consumer discrimination among lots in the market

Seed testing methods include:

- ❖ Germination test
- ❖ Viability test
- ❖ Purity test
- ❖ Weed test

Once the seed samples are received in the testing lab, the respective samples get registered and are assigned with laboratory test numbers and code for subsequent analysis. The procedure for seed testing involves:

- ❖ Conducting of moisture test, soon after registration of seed samples.
- ❖ Need of working samples from division of submitted sample for various tests using mechanical divider, random cups, spoon or by modified halving method.
- ❖ Calculating the weight of working samples for purity analyses with 2500 seeds or maximum of 1000 gm of sample weight.
- ❖ The working sample is separated into selected component and rest component, where fractions are weighed and weight of former sample is calculated as percentage of both.

Seed testing samples used in Seed Testing Laboratory includes:

- ❖ Service sample: It is a sample which is received from other than seed certification agencies and seed inspectors
- ❖ Certified sample: It is a sample which is received from certification agencies or officers
- ❖ Official sample: It is a sample which is received from seed inspectors.

1.4 Label and send seeds samples to lab technician as per company SOP

1.4.1 Seed Testing Laboratory

- ❖ Seed samples are inspected for presence of weeds, inert material and other quality factors like germination rate so that consumers/farmers know that, seeds they purchase is of highest quality.
- ❖ States and Union territories maintains seed laboratory in order to test different seed samples which were submitted by field representatives, seed companies and crop growers.
- ❖ Seed laboratory is authorized by law to test seed samples as a service to public by charging certain fees.
- ❖ In this, seed analysts and technicians perform the necessary seed testing functions.

1.4.2 Submission of Labelled Seed Sample

- ❖ Seed samples are tested for purity to determine the amount of pure seed, other crop seed, weed seed, noxious weed count, and inert matter that are present in the sample.
- ❖ The labelled seed sample should be submitted to lab technician with following details:

- Seed Type
- Name and address of company/individual
- Type of test requested
- Lot number
- Treatment Type(s)

2

Segregate Seed Bins for Pre-Cleaning

2.1 Keep seeds for pre-cleaning if undesirable material like dust, small stones, dirt balls, trash, leaves, pieces of stems, awns from seed heads, pieces of maize cobs, pieces of seed pods, dead insects etc are found in them

2.1.1 Pre-Cleaning of seeds

- When seed are brought in from the field, they contain undesirable materials such as pieces of stems, dust, weed seeds, other crop seed, and off-size, discolored, broken, and otherwise impaired units of the crop seed.

- Before the seed can be sold for planting through regular trade channels, enough of the undesirable materials must be removed to enable the seed to meet quality standards demanded by the farmers and required by law.
- Pre-cleaning is preparation of threshed seed for subsequent processing operations which involves winnowing, scalping and de-bearding.
- Pre-cleaned seed flows better and improves overall performance of processing plant.
- Precleaning is useful when there is large amount of trash or green, wet material in seed.

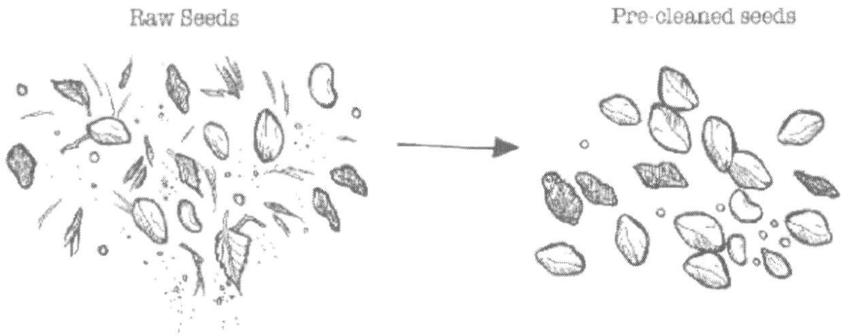

- Precleaning before drying can result in much enhanced capacity of dryers and promotes better flow through conveying equipment by allowing higher capacity and/or more precise screen selection in operation of cleaning equipment.
- The seed which will be graded or treated needs to be properly cleaned and conditioned by removing seed appendages and other materials from the seeds.
- Pre-cleaning of seeds on the farm is done to remove large unwanted materials like straw, large stones, etc.
- To pre-clean the seeds, grower/farmer can use equipments such as scalpers, debearders, huller-scarifier, Buckhorn machine and maize sheller..

2.1.2 Methods of Pre-cleaning of seeds

Winnowing

- It separates heavy and light material using a simple winnowing tray
- In this, the material from the tray is dropped through stream of wind on clean area.

Scalping

- It uses scalpers of different forms where common scalper has two screens which separates seed into three parts.
- In scalper, screens are fixed inclined which is then mechanically vibrate when raw seed passes through it.
- The raw seed is fed on top flat screen of scalper, which removes large debris, straw, leaves and allow seeds and other small particles to fall on the screen.
- The seed on second screen in scalper gets discharged in hopper, while smaller impurities fall through the holes of lower screen and are swept away by draft of air.

❖ The other type of scalper is round drum which is made from perforated metal which is set inclined in which, a drum rotates and seeds fall through the perforations; thereby removing debris in the drum.

De-bearding

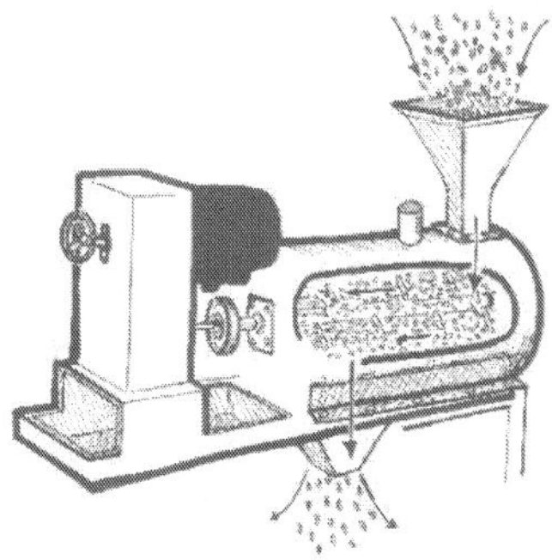

❖ In this, de-bearder/de-awner breaks clusters of material and removes awns, beards, tips or glumes on certain seeds to improve the flow of seed by cleaning machines.

❖ The de-bearder is installed in front of air-screen cleaners where a bypass mechanism for seeds does not need de-awning.

❖ The de-awner is enclosed in steel case equipped with shaft fitted with staggered steel beaters where shaft rotates and beaters breaks the material on passing through the machine.

❖ In this, the rotation speed of shaft is regulated which adjust the distance between the beaters and discharge shutter to limit the damage on seed.

2.1.3 Cleaning of pre-cleaned seeds

❖ When a grower/farmer have completed the pre cleaning process of seeds, he/she needs to clean the seeds again.

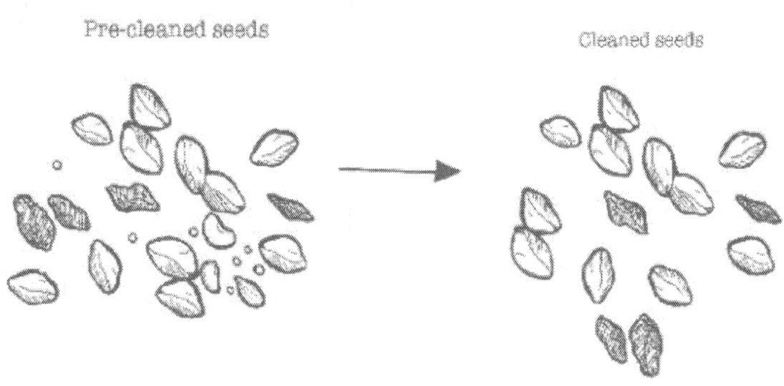

❖ Cleaning is an important operation which helps in upgrading physical quality of seeds.

- In basic cleaning, broken and rotten seeds are removed along with any impurities which have been missed during pre-cleaning process.
- For cleaning of seed, seed size and weight of seeds are considered in seed cleaner equipments at farm.
- The basic principles which are used in cleaning includes aspiration, scalping and grading.
- Grower/gardener can use air screen cleaner equipment for seed cleaning which functions similar to blower, where there are four levels, that are used to screen seeds of four different sizes and in separation.
- For good cleaning process of seeds, grower/farmer should check for moisture content of seed and should see for seed bins, elevators, seed cleaners and gunny bags.

2.2 Keep those seed bins, which are clean and does not require pre-cleaning, separately

2.2.1 Pre clean of seed bins

- Cleaning and preparing of seed bins will help seed processing worker to ensure that seeds/grain insects didn't diminish the value of harvest.
- Seed worker should clean the seed/grain bins thoroughly using vacuum cleaner by disposing of spilled, cracked and broken grain and grain flour, along with insects feeding on such material.
- If the seeds bins are pre-cleaned, it will:
    - decrease the infestation risk
    - improves sanitation and dust control
    - decreases microbial growth
    - improves flow of grain through the bin
    - increases storage life of grain

Seeds after pre-cleaned should be stored in pre-cleaned bins, so that there should not be any risk of infection and damage to seeds.

2.3 Send clean seeds for fine cleaning

2.3.1 Introduction

- As seed from basic cleaning process is reasonably clean and high in purity, still the process of fine cleaning is required to raise the physical purity to maximize standard possible in form of pure seed.
- Normally for fine cleaning of seeds, common machines like indented cylinders and gravity separators can be used.

- Seed processing worker after taking pre-cleaned seeds, should send the pre-cleaned seeds for fine cleaning which can be normally done using machines.

2.3.2 Machines for Fine cleaning

The seed processing worker after pre-cleaning of seeds, will allow the seeds for fine cleaning to obtain more purity in seeds.

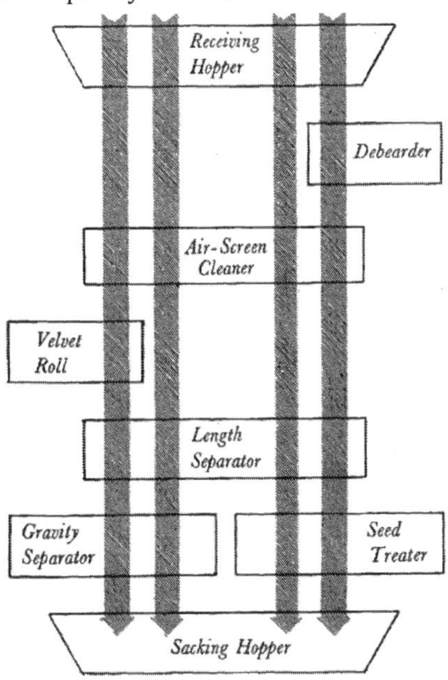

For this, seed processing operator uses following seed cleaning machines:

Indented Cylinder

- It grades the seed by length and helps in removing unwanted materials which are longer or shorter than required seed.
- The machine is installed as single unit or as several units in series or parallel as per the specific need of separation process.
- In this, cylinders are installed in tandem, where first cylinder removes long impurities while other cylinder removes short particles.
- The indented cylinder machine contains a cylindrical shell with imprinted pockets on inside surface, screw conveyor with shaft, trough, main drive arrangement and vent for connection to the dust collection system.
- The size of indented pockets on inside surface of cylinder depends on length of material to be lifted and removed.
- For holes in cleaner screens, even small differences in indents can have significant impact on quality of seed processing, so that wider range of different cylinders lead to betterment.

- ❖ For many crops, the main purpose of indented cylinder is to remove broken seeds.

Operation:

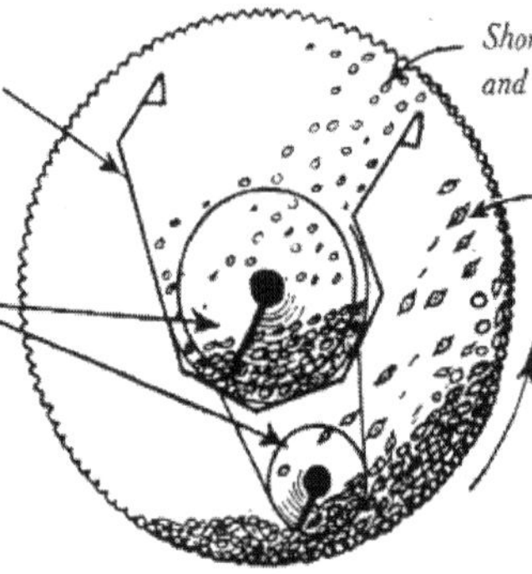

- ❖ In this, seeds are fed through the inlet of indented cylinder where seeds move through the rotating cylinder which helps in keeping smaller and broken seeds in pockets.
- ❖ While doing this, small seeds drops in collecting trough and are discharged by screw conveyor.
- ❖ In this, seeds larger than indent pockets will remain in the cylinder shell which are transported to the outlet and collected as final seed.
- ❖ During operation, the feed rate is adjusted as if feed rate is too low, then working capacity of cylinder is not optimized while if speed rate too high, then separation will result poor as individual seeds may not enter the indents for sizing, so feed rate should be uniform to optimize the capacity.
- ❖ While cleaning, the machine operator must take random samples from discharge spouts and should examine the seeds in order to check the quality of separation.

Gravity Separator

- ❖ The gravity separator separates seeds of similar size but different specific weight and grades the cleaned seeds into quality fractions as per specific weight.
- ❖ It is a critical part of processing line for high-value crops while for cereal seeds that have been through basic cleaning and fine cleaning with indented cylinder, the machine gives very little return on an eventual investment.
- ❖ It contains an inclined deck made of special square wire which is meshed mounted on sturdy support frame.
- ❖ In this, the deck can be rectangular or triangular in shape is fitted with fully balanced eccentric drive that causes deck to oscillate.

- For light seeds, decks are made of special cloth.
- All kinds of deck separates lighter and heavier components of seed lot, while triangular models remove lighter seed more efficiently but are also good for removing heavy seed.
- In the machine, below the deck is a fan with a pressurized air supply system.
- As per type of seeds, angle of inclination of deck, oscillation speed, quantity of air and pressure can be adjusted.

Operation:

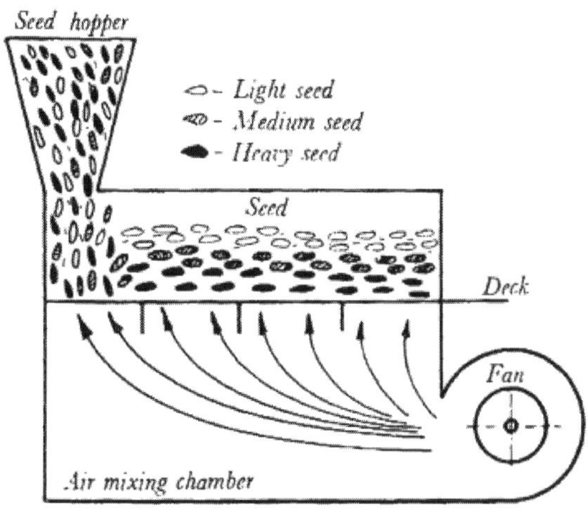

- In this, seeds are fed continuously onto the deck in steady stream to give uniform spread over the deck surface.
- The pressurized air from below the deck causes the bed of seed to fluidize.
- Here, light seeds float to the top while heavy seeds remain in contact with wire mesh surface.
- In this, the deck oscillates which moves heavier seeds uphill where the seeds get discharged through various collecting spouts fitted at the end of deck.
- The separation process results in spectrum of seeds from light to heavy where practice and experience is required to achieve high-quality separation, as operators need to do certain adjustments.

Spiral separator

- The spiral separator was developed to remove unwanted particles from seeds as per the ability to roll.
- The machine comprises of column with spiral slide that has double bottom where seeds are loaded on top of slide to roll down.

- The perfectly spherical seeds in this, rolls down faster and reaches at the edge of slide which falls into separating trough while non-spherical contaminants rolls more slowly and remain in inner slide.
- The spiral separator separates seed of highest quality, but it is only suited to species with spherical seeds.
- When it is used in a seed-processing plant, groups of spiral separators tend to operate separately from the rest of the plant.

Velvet roller

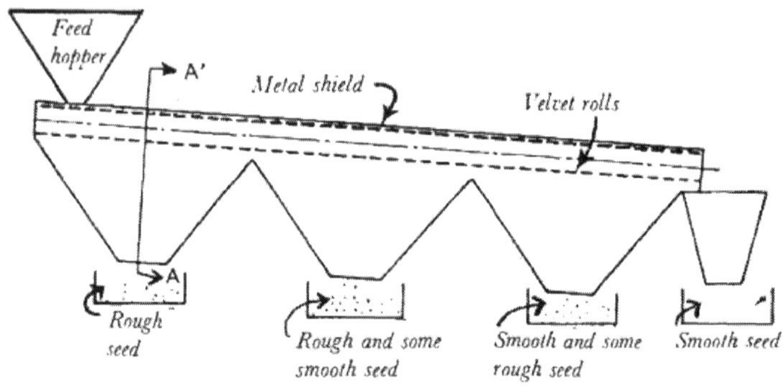

- The velvet roller separates seeds of same size but differing in roughness.
- It is used for forage species due to variation in flowering patterns as seed maturity is never homogeneous at harvest and many seeds become wrinkled once dried.
- The machine comprises of two parallel cylinders that are attached together and covered by a special cloth made of natural fibre.
- In this, cylinders are slightly inclined and rotate in opposite directions at 0.5-1 revolutions per second.
- The seed is placed between two rotating cylinders which moves inclined down.
- In this, rotational movement lifts the seeds with sharp edges or rough surfaces which falls in collecting bin which separates seeds with different degrees of surface roughness.

3

Perform Pre-cleaning/ Fine-cleaning of Seeds

3.1 Remove dust, debris, trash etc. from un-cleaned seeds using graded sieves as per SOP

3.1.1 Introduction

- Cleaning seeds is important as it removes unwanted materials from it. A clean seed has high value than contaminated seeds having straw, chaff, weed seeds, soil, rubbish, and other non-grain materials.
- Malformed, discolored, germinated, broken or moldy grains in seed lots can severely impact seed quality, viability and vigor.
- In many cases, number of sieves with different sized perforations are used and cleaning sifts out smaller and smaller particles.
- It is not only the size of perforations which shows quality and quantity of seed cleaned, but certain factors include precision of perforations, angle at which sieves operates, amplitude and speed of movement of sieves and correct cleaning and maintenance of equipment.
- Sieves are made of flat perforated plate or wire mesh which are three dimensional like funnel shaped.
- For small samples, hand held sieves are adequate, but for larger scale cleaning, series of shaking screens is commonly used.

3.1.2 Remove dust/debris/trash from seeds using graded sieves

Sieves are made of flat perforated plate or wire mesh which are three dimensional like funnel shaped. Following are the graded sieve machines used for removing dust, debris, trash etc from seeds which includes:

Machine 1:

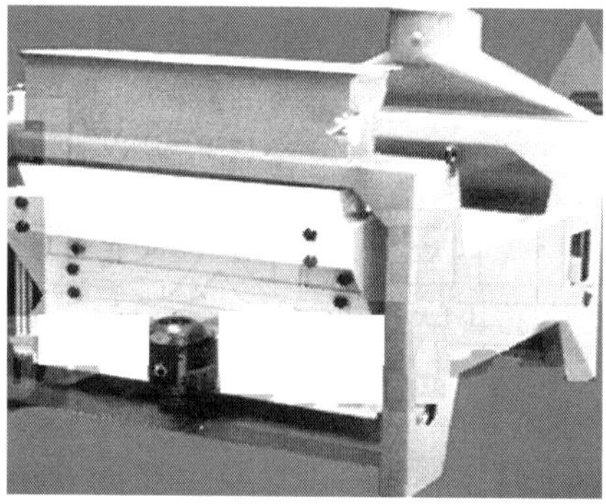

- ❖ The graded sieve machine is used for cleaning/ grading of oilseeds using set of two sieves, to remove coarse and fine impurities and attachment to aspiration system which is used for removing low-density matter.
- ❖ The machine comprises of frame structure which is made of steel channel with fully enclosed sieve box that is suspended at its four corners on rubber bushes.
- ❖ In this, box has two decks with ball cleaned sieves and vibrator drive is located in center.
- ❖ The vibratory movement of sieve increases the capacity on a given sieve-surface and sieve effect.
- ❖ The capacity of the machine ranges from 1-10 TPH
- ❖ The machine is used for thorough removal of dust, sand, stones or unwanted particles and is easy to operate.

Machine 2:

- ❖ The graded sieve machine is suitable for cleaning of all types of grains and seeds.
- ❖ It is used to separate out impurities, which are bigger/smaller than seed size.
- ❖ The capacity of the machine ranges from 1-3 TPH

- The machine is used for thorough removal of dust, sand, stones or unwanted particles and is easy to operate
- The machine enhances the life of spares of oil expellers where magnet placed at discharge end is used for removal of iron pieces

Machine 3:

- The graded sieve machine is a type of rotary seed cleaner which is suitable for cleaning of all types of grains and seeds.
- It is used to separate out impurities, which are bigger/smaller than seed size.
- The capacity of the machine ranges from 1-3 TPH
- The machine is used for thorough removal of dust, sand, stones or unwanted particles and is easy to operate

Vibrating Separator

- The graded sieve machine is suitable for efficient cleaning of all types of grains and seeds.
- It is used to separate out impurities, which are bigger/smaller than seed size.
- The machine is fabricated from Mild Steel Plate, Sheet, Angle and Channel.
- It has self-cleaning rubber balls on bottom sieve.

- In this, the intake hopper of machine is equipped with canvass clothes.
- The machine has two vibrating screens, where top screen is used for removing over size impurities while bottom screen is used for removing under size impurities like sand, shriveled grain etc.
- The machine is driven with suitable two vibro electric motors.

3.2 Separate lightweight material and empty glumes by gentle winnowing

The glumes or husks of grains are separated from seed by gentle winnowing.

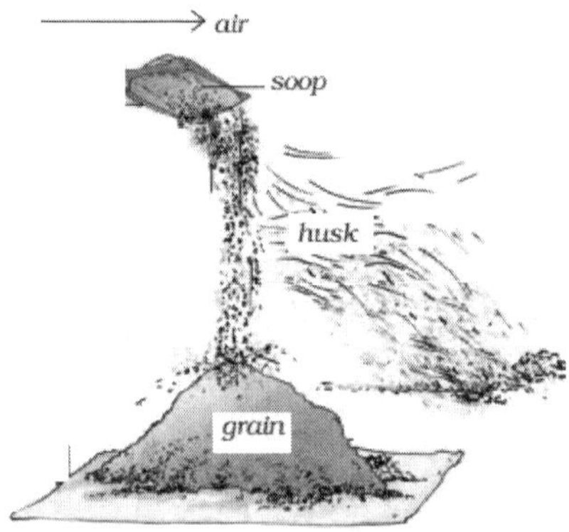

- It is a method which involves throwing the mixture in air so that the wind blows away the lighter chaff, while heavier grains fall back down for recovery.
- It is used to separate heavier and lighter components of mixture using wind or by blowing air.
- It is commonly used by seed growers/farmers to separate lighter husk particles from heavier seeds of grain.
- In this, husk particles are carried away by the wind and the seeds of grain gets separated and form heap near the platform.

3.3 Remove undesirable materials using the seed blower as per SOP

- Once the seed is ready after conditioning and pre-cleaning and after removing inert material, weed seeds, other crop and broken seeds which are big or small than normal seed, then crop processing worker needs to separate the undesirable materials from the seeds.
- Seed blower makes separation of undesired materials accurately by controlling upward air flow by transparent plastic column for purity testing, docking and cleaning of grain/seed samples.

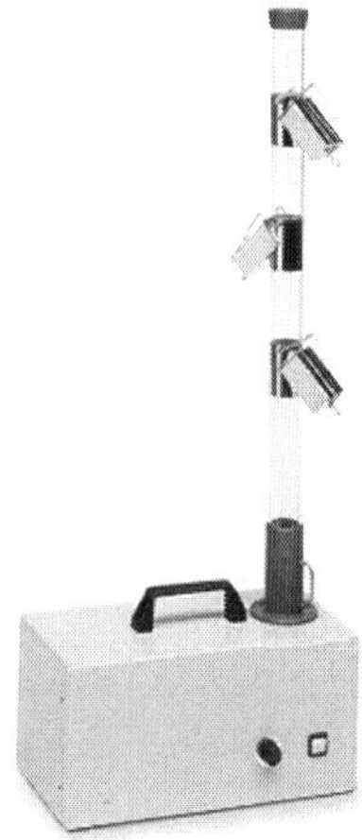

- It serves as quick and easy separation of light and heavy seed.
- In seed blower, air flow can be controlled to get most difficult sensitive air separations.
- In this, seeds are filled in drawer of glass cylinder and fine-meshed sieve closes the upper end of cylinder.
- A speed-controlled blower in cylinder, generates constant upward air column.
- As per the size and weight, seed will fall in one of collecting vessels being installed at cylinder where light seed gets deposited in upper collecting vessel while heavy seeds will be deposited in lower collecting vessel.
- In this, sieve at the bottom of cylinder avoids penetration of seed in blower.
- Once blower is disconnected, sieve contains deposit of residues and of small stones.
- The collecting vessels are equipped with snap buckle clip which allows easy removal.
- Spread seeds on a flat well-lit surface of contrasting colour such as an illuminated table to remove damaged seeds, seeds of different species etc. if any

3.4 Send seeds for fine cleaning

- As the seeds from basic cleaning process are almost clean and good in physical purity, still the seed processing worker needs to send the cleaned seeds for pre-cleaning.
- The processes of fine cleaning and grading are required to raise the physical purity to maximum standard possible in form of pure seed.
- Specialized machines separate the seeds as per density, length, width, thickness and shape.

- Fine cleaning improves seed quality of cleaned seed and helps in removal of oversized and undersized seeds and other particles like stones and sand where cleaned seed goes in outlet for bagging or for further processing.
- Seed grading achieves highest possible level of purity and normally targets specific contaminants.

3.5 Clean seeds further by using 'Fine Cleaning Machine' as per the SOP

3.5.1 Introduction

The machines which are commonly used for fine cleaning of seeds are indented cylinders and gravity separators, which are placed after the air-screen cleaner.

3.5.2 Indented Cylinder

- It grades the seed by length and helps in removing unwanted materials which are longer or shorter than required seed.
- The machine is installed as single unit or as several units in series or parallel as per the specific need of separation process.
- In this, cylinders are installed in tandem, where first cylinder removes long impurities while other cylinder removes short particles.
- The indented cylinder machine contains a cylindrical shell with imprinted pockets on inside surface, screw conveyor with shaft, trough, main drive arrangement and vent for connection to the dust collection system.
- The size of indented pockets on inside surface of cylinder depends on length of material to be lifted and removed.
- For holes in cleaner screens, even small differences in indents can have significant impact on quality of seed processing, so that wider range of different cylinders lead to betterment.

- For many crops, the main purpose of indented cylinder is to remove broken seeds.

Operation:

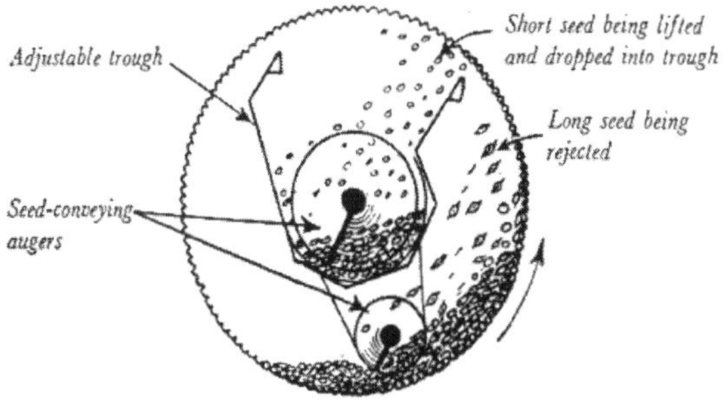

- In this, seeds are fed through the inlet of indented cylinder where seeds move through the rotating cylinder which helps in keeping smaller and broken seeds in pockets.
- While doing this, small seeds drops in collecting trough and are discharged by screw conveyor.
- In this, seeds larger than indent pockets will remain in the cylinder shell which are transported to the outlet and collected as final seed.
- During operation, the feed rate is adjusted as if feed rate is too low, then working capacity of cylinder is not optimized while if speed rate too high, then separation will result poor as individual seeds may not enter the indents for sizing, so feed rate should be uniform to optimize the capacity.
- While cleaning, the machine operator must take random samples from discharge spouts and should examine the seeds in order to check the quality of separation.

3.5.3 Gravity Separator

- ❖ The gravity separator separates seeds of similar size but different specific weight and grades the cleaned seeds into quality fractions as per specific weight.
- ❖ It is a critical part of processing line for high-value crops while for cereal seeds that have been through basic cleaning and fine cleaning with indented cylinder, the machine gives very little return on an eventual investment.
- ❖ It contains an inclined deck made of special square wire which is meshed mounted on sturdy support frame.
- ❖ In this, the deck can be rectangular or triangular in shape is fitted with fully balanced eccentric drive that causes deck to oscillate.
- ❖ For light seeds, decks are made of special cloth.
- ❖ All kinds of deck separates lighter and heavier components of seed lot, while triangular models remove lighter seed more efficiently but are also good for removing heavy seed.
- ❖ In the machine, below the deck is a fan with a pressurized air supply system.
- ❖ As per type of seeds, angle of inclination of deck, oscillation speed, quantity of air and pressure can be adjusted.

Operation:

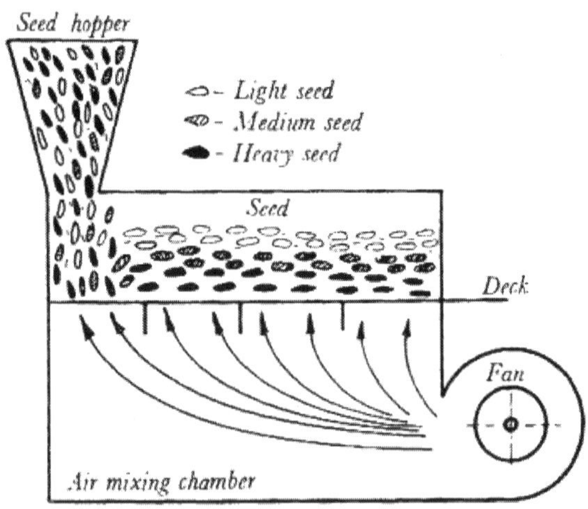

- ❖ In this, seeds are fed continuously onto the deck in steady stream to give uniform spread over the deck surface.
- ❖ The pressurized air from below the deck causes the bed of seed to fluidize.
- ❖ Here, light seeds float to the top while heavy seeds remain in contact with wire mesh surface.

- In this, the deck oscillates which moves heavier seeds uphill where the seeds get discharged through various collecting spouts fitted at the end of deck.
- The separation process results in spectrum of seeds from light to heavy where practice and experience is required to achieve high-quality separation, as operators need to do certain adjustments.

3.6 Clean seeds further by using 'Air Screen Cleaner' as per the SOP

3.6.1 Introduction

- Once the initial cleaning of seed has been done, the second stage of seed cleaning done using air blasts and vibrating screens with machine known as air-screen cleaner.
- The air-screen machine is basic cleaner used in many seed processing plants where almost every seeds are cleaned by air-screen cleaner before specific specifications can be attempted.
- Air/screen seed cleaners are the heart of any seed conditioning facility and are such type of cleaning equipment which initially exposed to incoming seed.
- Proper cleaning of seeds by air/screen unit is must to allow other finishing types of equipment to make separations for which they were designed.
- Air-screen machine effects separations based on differences in size and weight of seeds.
- The cleaner separates seed from contaminants by use of aspiration and screens of perforated metal or wire cloth which are present in wide range of sizes.
- Operation of air/screen cleaner needs to be fully understood by seed machine operator as it will help him/her to make best possible separations for seed being cleaned.

3.6.2 Construction /Working

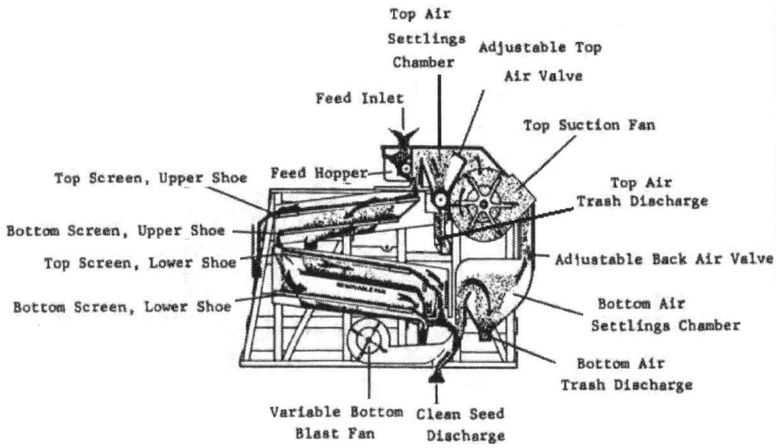

- Air/screen seed cleaners has top and bottom air separations which helps the screens in cleaning seed mass.
- The top air separation is made as seed passes from feed hopper to top screen in upper shoe.
- Air is pulled by seed mass lifting light trash, stems, pods, straw, cockleburs and part of broken seed/splits which are dropped in settlings chamber and will discharged out the side of cleaner by spout/discharge auger.
- The top air needs to be set to pull as much of light contaminants as possible.
- A proper setting will help in removing one or two good seed per double handfull of screenings to allow operator to know that he/she is getting the best possible air separation without losing extra amount of good seed.
- The bottom air separations are made after screen cleaning and before seeds discharge from the cleaner.
- The bottom air lifts any light trash that may be left in seed commodity.
- While cleaning, cleaner with good bottom fan, properly adjusted, will help in removing all slick cocklebурs that may be left in seed mass after screen cleaning.
- With top air separation, proper setting for bottom fan needs to pulled out occasionally good seed.
- If bottom fan works in conjunction with top fan, then two fans needs to balance for most effective and efficient separation.
- The top fan should pull slightly more cubic feet of air per minute than bottom fan and proper balance can be set usually with use of ribbon indicator.

- In this, the hopper is a feed mechanism and not a surge bin as cleaners normally works well and efficient with surge bin above hopper to give steady supply of seed to hopper portion of cleaner.
- The discharge gate which causes hopper to be center-fed where overhead surge bin to be of same width and length as feed opening in cleaner hopper.
- The center-fed hopper will feed incoming seed mass to center part of top screen which causes cleaner to lose one-third of its available perforated screen area.
- In the figure, common seed flow over precision air/screen cleaners with four screens is SCALP-SIFT, SCALP-SIFT system, which shows seed cleaning setup.
- The three scalping screens are used when seed mass to be cleaned gets heavily laden with or with large foreign materials having many scalping with perforations of different sizes with small parts of broken, weed seeds and sand.

4

Dry the Seeds

4.1 Receive test report from the lab technician

4.1.1 Introduction

- In seed testing, initial step in managing seed testing laboratory starts with seed sampling management to get sample of size suitable for seed tests.
- In this, seed lot to be sampled has to be of uniform in composition and size so that the sampler should have access to each and every container.
- Once seed testing is done in lab, the test results are entered on analysis cards for lab technician to prepare seed analysis report.

4.1.2 Seed test report

- Once the seed sample is tested in the laboratory, the test report will be handed over to seed processing worker by seed technician.
- The tested seed report shows different seed characteristics.
- If seed growers understand the significance of different classifications on their test report, they will be able to understand, what will be the opportunity for that seed crop.
- In a test sample, if seeds are abnormal, it will result in number of different factors like frost damage, heating in bin or seed-borne diseases.
- Seed testing labs also put the methods and procedures they have used in their test report such as temperature at which germination test was conducted which gives indication about growing conditions, the seed will respond to.

4.2 Checking moisture level and send the seeds for further processing

4.2.1 Introduction

- Seed samples needs to be securely contained in moisture proof container for moisture test and should be delivered to laboratory.

- ❖ Seed laboratories after receiving the seed sample will conduct the sample test for seed quality attributes to meet the needs of seed law enforcement, seed certification and service samples as received from seed growers.
- ❖ Seed tests are done for checking seed moisture, physical purity, germination and seed health as per seed certification standards.

4.2.2 Seed moisture

- ❖ The moisture content is amount of water in the seed which is expressed as percentage.
- ❖ The moisture inside seed can be expressed on net weight basis or dry weight basis.
- ❖ The moisture content in seed can be expressed on dry weight basis in db, where loss in weight as percentage of dry weight of seeds, so calculating seed moisture content(%db) = (wet weight – dry weight)/dry weight x 100.
- ❖ Seed moisture content can be accurately determined experimentally by scientific techniques, or can be predicted from information.
- ❖ The determination is destructive to seeds used, as in many cases, prediction is sufficient, so determination should only be done where required.
- ❖ The moisture content inside the seed has large effect on storage life of seeds, hence it is important to know the moisture content in order to make an accurate prediction about storage life of every accession.
- ❖ Moisture in seeds has strong bearing on length of time till it remains viable.
- ❖ Seeds may sprout or molds develops at high levels of moisture and seeds tends to lose viability in few days.
- ❖ At ordinary temperatures, if relative humidity of air around the seeds is more than 75%, then seeds support the growth of molds to an extent that they should not be stored even for shorter time.
- ❖ The moisture content of seeds in equilibrium with critical relative humidity varies among different kinds while moisture levels below those causes actual sprouting or mold development high to support fairly active physiological activity in living seeds.
- ❖ The optimum moisture level for storage of many kinds appears to be between 6 and 8 percent.

4.2.3 Predicting seed moisture content

For predicting the seed moisture content from test report, seed processing worker needs to follow:

Measurement of environmental conditions

- Measure relative humidity of atmosphere where the seeds have been held for past few days using aspirated hygrometer.
- Measure mean temperature where the seeds have been held for past few days.
- If seeds are from different environment, they need to be allowed to equilibrate with current one before making any predictions.

Checking equilibrium moisture content

- Obtain the name of crop in left hand side of tables of equilibrium moisture contents.
- Read equilibrium seed moisture content at relative humidity measured as shown above.
- Use the figure of approximate moisture content, since predicted value in making estimates of drying period or exact value of moisture content is not necessary.

Procedure

Seed processing worker needs to follow following procedure to test for seed moisture which includes:

Sample Collection:

- Collect moisture samples immediately on arrival at sample reception.
- Ensure that the sample has a test number.
- The moisture samples are taken to moisture room where they are arranged in order of test numbers.
- Care should be taken to avoid moisture gain or loss.
- The sample should be mixed thoroughly as prescribed by standards and rules.
- A moisture sample shall be tested as soon as possible, within 24 hours of receipt at laboratory

Checking status of moisture meter:

- Clean the working surface area and ensure that moisture meter is free from previous analysis.
- Check the power supply of the meter
- Switch ON moisture meter

Preparing sample:

- Prepare a duplicate of two independent working samples as per current version of rules.

- ❖ Ensure that the duplicate sample is not exposed to atmosphere for more than 30 seconds.

Loading sample:

- ❖ Pour the first duplicate sample to fill the hopper of moisture meter and compress sample as per instructions manual of moisture meter.
- ❖ Record the measurement on the analysis card.
- ❖ Now pour second duplicate sample to fill the hopper of moisture meter and compress sample as per instructions manual of moisture meter.
- ❖ Record another measurement on analysis card.

Calculating results:

- ❖ Calculate the mean moisture content using observed percentage of two samples to decimal place by formula (Sample1+Sample2)/2, Where Sample1= moisture content of first duplicate sample while Sample2= moisture content of second duplicate sample
- ❖ Now express the result as percentage to one decimal place.
- ❖ Interpreting results:
- ❖ If the difference between replicates does not exceed 0.2%, then arithmetic mean is taken as result.
- ❖ In the event that 0.2% exceeds, then do a duplicate test.

Oven-drying method

- ❖ The accurate method for determining moisture content in seeds is by using oven-drying method, where water is removed from seeds by heat under controlled conditions.
- ❖ The method is destructive to seeds and should be carried out only when essential.
- ❖ It is recommended that one accurate determination be conducted using such method after drying period to determine initial moisture content of stored seeds.
- ❖ It uses following equipment in determining moisture content:
    - mechanical-convection oven with recovery time of 15 minutes or less to maintain required temperature within 1°C and fitted with thermometer accurate to 0.5°C
    - non-corrosive drying containers with tight fitting lids and size of container should allow the height of evenly-distributed sample to be under 0.3 g/cm^2
    - grinder with adjustable speeds to get specified particle sizes of 0.5mm–4.0mm which should not cause undue heating while grinding

- analytical balance which is capable of weighing to 3–4 decimal places as 0.001g–0.0001g
- desiccator fitted internally with thick metal or ceramic plate for rapid cooling of containers and containing desiccant like silica gel or calcium chloride at bottom
- tongs or gloves to handle hot containers

Seed moisture during harvesting of seed crops

Seed moisture during harvesting

- Seed moisture content is most reliable indicator of seed maturity and harvest timing in seed crops.
- There are significant times during harvest as knowledge of seed moisture is critical: at swathing and combining.
- Swathing in correct range of seed moisture content maximizes seed yield and minimize seed losses during the harvest.
- Taking accurate measurement of seed moisture content is main component of economic seed crop management.

Seed moisture during threshing of seed crops

Threshing

- Threshing detach the grain from panicles which can be done by rubbing, stripping, impact action or using all actions.
- Threshing operation can be done manually (trampling, beating) using animal power or mechanical threshers.
- Manual threshing is common practice in developing countries.
- Grain spillage, incomplete separation of grain from chaff, grain breakage due to excessive striking, are main causes for loss during threshing process.
- Thresher, which is a machine that separates grain from harvested crop provides clean grain without much loss and damage.
- In threshing, seed loss due to broken seed, unthreshed seed, blown seed, spilled seed etc. can be minimum.
- Clean unbruised seed fetches good price in the market and has long storage life.
- Threshing by bullock treading is practiced on large scale in the country but it is also time consuming and involves drudgery. Tractor in many places is now used in place of animals for treading.
- Methods of threshing influence seed quality in terms of germination and vigour.

- The different threshing methods produces breaks, cracks, bruises and abrasions in seeds which in turn results in abnormal seedlings of questionable planting value.

Impact of threshing and moisture content on seed quality

- The required threshing moisture for certain seeds should be around 14%-16%, so threshing above moisture content will show deterioration in seed quality.
- The mechanical damage to seed during threshing varies as per genetics of variety.
- Seed hardness plays an important role in seed processing and vitreous endosperm variety pose breakage resistance.
- The percentage of damaged seed using wooden stick was higher and different from threshed using paddy thresher.
- The percentage of seed germination was higher using paddy thresher and different from threshed using wooden stick.

4.3 If moisture level in seeds is higher than acceptable level, then send the seeds to seed drying area / room

4.3.1 Analyse moisture content in seed

- Moisture content is important factor in determining the rate at which seeds deteriorate and has profound impacts on storage longevity of seeds in gene banks.
- Small changes in moisture content have large effects on storage life, so it is important to determine moisture content before storage in order to accurately predict the potential storage life of each accession.
- Moisture content can be expressed on dry weight basis in db, where loss in weight as percentage of dry weight of seeds as Seed moisture content(%db) = (wet weight – dry weight)/dry weight x 100.
- The relationships of seed moisture content on wet weight basis to seed moisture content on dry weight basis, and of equilibrium moisture content of seeds to relative humidity of surrounding atmosphere.
- In old seeds, moisture content is an important single factor in determining seed longevity.
- Reduction in moisture content causes reduction in respiration which slows down ageing of seed and prolongs viability.
- There are different seed moisture content% (wet weight) which describes different seed behaviour which shows that if seed moisture content percentage is:

- above 45% – 60%, the germination of seed starts
- above 18% - 20%, then seed may be heated due to rapid rate of respiration and energy release
- above 12% – 14%, then fungus growth in seeds takes place
- below 8% – 9%, then insect activity will get reduced
- below 4% – 8%, then seed can be stored safely in sealed place/container

- It is observed that fluctuation in seed moisture content while storage due to open storage without humidity control or frequent opening and resealing of sealed containers will results in deterioration in germinability of seeds.
- It is observed that steady moisture content slightly above the optimum value is normally less harmful as compared to one which fluctuates between the optimum and higher moisture content.
- Moisture content is important in recalcitrant seeds, but in such case, critical moisture content is minimum to which it is allowable to dry the seeds rather than maximum content for prolonged storage.
- For large hardwood seeds, moisture contents is appropriate in range of 25% – 79%.
- Seeds at high moisture level can be stored for longer periods at freezing temperatures than at higher temperatures, while higher storage temperatures of 30°C are less harmful when moisture content of seed is low.
- The critical moisture content lies at high level when storage temperatures are low than when they are intermediate or high, so low temperature can compensate for high moisture content and vice versa.

4.3.2 Seeds for drying

- It is required to dry moist seeds before processing and storing as seeds with high moisture content are more susceptible to physical damage during processing.
- Immediately after harvest, moisture test shows if further drying is necessary.
- As life of seed mainly moves around its moisture content, so it is necessary to dry the seeds to safe moisture contents.
- The sage moisture content will depends on storage length, type of storage structure, kind / variety of seed type of packing material used.
- For cereals in ordinary storage conditions for 12-18 months, seed drying till 10% of moisture content will be quite satisfactory, but storing of seeds in sealed containers and drying upto 5-8 % moisture content will only depend on particular kind which is required.
- Drying is post-harvest system phase where product is rapidly dried till it reaches safe-moisture level.

- The idea of drying the seed grains is to lower the moisture content in it for safe storage and further processing.
- Drying in growing countries can be done by spreading thin layer of seeds in sun, on the threshing floor or on rooftops.
- Seed moisture content is one of the factors which determine whether or not seed can be stored safely without loss of germination and vigour.
- When moisture content is too high, seed may heat and various moulds can grow, so it is vital to ensure that harvested seed is at safe moisture content before putting it into store.
- Safe seed moisture contents vary with crop species, as 14% or less is considered satisfactory for short term storage.

4.4 Moist seed in Seed Drying Area

- Drying is intended to reduce moisture content of seeds at required level, so that post harvest operations can be done, and seeds can be kept in safe storage till their marketing and seeding.
- Drying prolongs storage life, allows seeds to tolerate extreme temperatures, prevents germination, and reduces attack by predators and pathogens
- The reduction of moisture decreases available water, altering the metabolism and slowing down all the biochemical reactions that decrease the physiological seed quality, thus providing a safe and long storage.
- Drying of moist seeds is done in two stages where first drying is done by delaying the harvesting of seed crop while second drying involves windrowed or bundled of harvested seed crop which is left in the field or transported to threshing site and stacked for further drying.

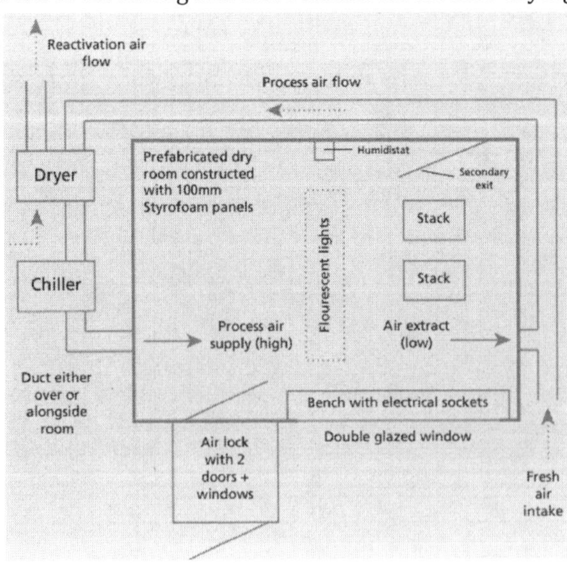

- Dry room is the most appropriate drying method when you need to dry large quantities of seed at the same time.
- Dry rooms also provide safe storage prior to seed cleaning and a suitable location to package samples for storage or to warm up frozen samples that have been removed from the cold room.
- The seeds drying room should be sealed with low relative humidity and cool temperature.
- The seed drying room should have good insulation to reduce energy usage while cooling and drying.
- The room can be constructed using prefabricated structure, either as stand-alone room or in larger existing room with construct walls and ceiling of styrofoam panels.
- The drying room should have drier which can deliver about six air changes per hour which creates heat during drying process and should ideally be placed outside the room.
- The process air supplied by chiller dries the room which is high up at one end of room and extracted low down at other end of room.
- In drying room, drying agent picks up moisture from process air which is re-dried by another air flow where air is drawn in dryer, heated, and passed over drying agent.
- The controlling humidistat is kept halfway along one of longer walls of the room
- In this, chiller needs to be kept in the room which cools the heat load from the dryer to get room temperature of around 15ºC.

4.5 Undertake drying of seeds/traditional or sun drying and mechanical or artificial drying depending upon available resources, type of seed, seed quantity and work instructions in the SOP

4.5.1 Seed Drying

- Drying reduces moisture content of seed to required levels for seed processing and storage.
- Drying seeds not only increases longevity in storage, but also prevents attack by pests and pathogens.
- Seed drying depends on climate at harvest as rapid drying will harm the seed as water gets withdrawn quickly or due to high temperature while slow drying results in effect of maintaining seed at high moisture content and relatively high temperature which accelerate deterioration that drying is intended to prevent.

4.5.2 Methods

For drying seed, normally two methods are used:

- natural drying
- artificial drying

1. Natural Drying

- In this, seed grains are exposed to air in sun or shade.
- For obtaining desired moisture content, seed grains are spread in thin layers on drying-floor, where it is exposed to air.
- In this, the duration depends on moisture content as required for safe storage.
- To get uniform drying, seed grains must be stirred frequently, especially if it is in direct sunlight.
- For effective drying, relative humidity of ambient air should not be more than 70%.
- Here, grains should not be exposed at night and the method should not be used in humid regions or during the rainy Seasons.
- The insufficient or excessively slow drying can led to severe losses in product during storage from self generated heat of green seed grain.
- Seed grains should be sun-dried in the middle of the day only and should not be dried in bags on a concrete floor.
- The process requires very low cost and has flexible capacity of grains o be dried up.
- It is a slow drying process which normally depends on the weather conditions.
- In this process, high risk of contamination with possible loss and risk due to birds and rodents.

Sun drying

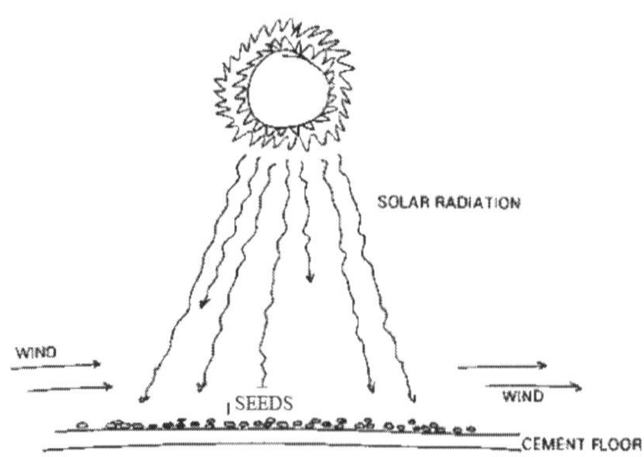

- Sun drying of seeds is drying method which is done when crop/seed is ready for harvest.
- Sun drying is traditional drying method for reducing moisture content which can be done by spreading seeds/grains under the sun.
- In this, solar radiations heats up the seeds/grains and surrounding air increases the rate of water evaporation from the seeds/grains.
- Some seeds like maize can be sun dried although, during drying, seeds/crops get sensitive to insect infestation, rodent or birds' attack, and mold damage.
- A threshed seed drying by spreading on sheets or tray is common phenomenon but has a risk of soil or stone contamination.
- The seeds are dispersed in thin layer and turned at regular intervals to facilitate drying and covered at night with the help of sheets.
- It is a common drying method as it needs low cost as compared to mechanical drying and is environmentally friendly as it uses sun as heat source where no carbondioxide is produced.
- For improving sun drying process, seeds needs to be spread in thin layers of 2–4 cm as very thin layers tend to heat up very quickly with negative effect on head seed recovery, so if layers are too thick, large moisture gradient develops with dry grains on top and wet grains on bottom, which will absorbs the moisture again after mixing, that lead to seed cracking.
- In good weather, seed processing worker should mix or turn the seeds in order to maintain good quality as variation in moisture content in seeds can lead to rewetting and subsequent cracking of dry seeds.

Limitations

- The drying directly depends on weather conditions and cannot be practiced during the night or rain
- High potential losses due to rodents, birds, insect pests, and diseases
- It has low efficiency; slow and labor-intensive method of drying
- In this, difference in temperatures and humidity during day and night affects the final success of drying as too long exposure to high temperatures may result harmful to some crops

2. Artificial drying

- In artificial drying, the vapour pressure gradient between the seed and air needed for drying can be obtained either by heating the air to increase its capacity to absorb water or by removing water from air.
- In this, heated or unheated air is passed through the seeds to remove moisture.

- In unheated drying, air has low relative humidity which is not effective in moist weather.
- With heated drying, it is faster and quicker to dry seeds as it needs less drying space.
- The natural drying does not deliver quality product in quantity due to uneven maturation of seeds in plant and among plants, with adverse weather conditions, hence forced air drying is essential to get high quality product, and such dryers are used in modern designs, which retain the product characteristics and good energy utilization.
- In this, heated air (dryers) or unheated air (dehumidifiers) is blown through grain mass.
- Such type of drying process has high initial costs and can cause damage of fire and possibility of overheating to seed.
- Mechanical drying techniques may be used with assistance of mechanical, electric or electronic equipment where air is forced by seed mass.
- Mechanical drying techniques afford the maintenance of seed quality, when conducted under technical criteria that define exposure times and temperatures.

Forced hot-air Drying

- This is the most wide spread practice in semi-humid and humid conditions where natural drying cannot be used.
- Artificially heated air is forced to flow through a mass of seed grain in bulk or in bags to absorb released moisture from grain mass
- It facilitates early harvesting and reduces shattering losses in the field
- It is more rapid drying in terms of capacity and is independent of weather conditions
- It requires heavy initial cost with high cost of drying
- In this, there is risk of seed damage due to high temperature
- It needs skilled personnel and regular monitoring with risk of breakdown or fuel shortage at critical times.

Dehumidified air-drying

- In this, unheated dehumidified air is circulated through seed grain mass till moisture content of grain is reduced to required level.
- It is commonly applied when small quantities of seed grains are dried to very low moisture content for being used as seed.
- It is simple and easy to manipulate which involves low risk of seed damage from hot air

- ❖ It is not good for large-scale production as it involves high cost of investment

Ideal drying-Intermittent dryer

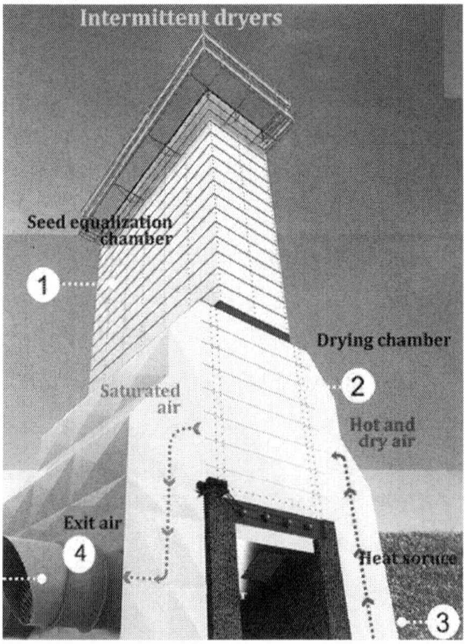

- ❖ Soon after physiological maturity, seed still attached to parent plant which begins lossing of water or natural drying.
- ❖ Such process occurs in nature so that the seeds are dispersed more easily by their vectors as birds or winds, and to maintain their viability to find an ideal place to spread.
- ❖ In cultivated species, between the point of physiological maturity and harvest time, seed moisture content should be enough to ensure product quality and to allow mechanical harvesting.
- ❖ In such sense, it is required to know the moisture content before knowing the drying concepts.
- ❖ The water is classified as fixed in seeds where:
- ❖ first type or monomolecular water content lies between 0%-5%, which is being strongly absorbed and closely linked to molecules.
- ❖ second type or polimolecular water lies between 5%-12% moisture and is strongly absorbed with no solvent function, but is no longer connected directly to molecules, but molecular chains.
- ❖ third type or osmotic has connection based on osmotic cell pressure, with solvent function solvent, and is poorly absorbed water, having its contents between 12%-27%.

- fourth type or free water which is being adsorbed and solvent water having more than 27% moisture content, can be easily removed.
- In this, free water is first to be removed by drying while third type is osmotic which will be removed with more difficulty where water content maintains seed quality.
- During such stage of water removal from seeds, seed grower should identify and select methods and drying processes, which do not result in compromise of seeds.
- For drying to occur, it is necessary that there is vapor pressure gradient between seed water (larger) and surrounding air (minor) which is obtained by reducing air relative humidity by heating or by dehumidification.
- Drying air should also have air flow to transport heat and water.
- In such process, there is energy transfer from air to evaporate surface moisture of seed, which depends on temperature, relative humidity, air flow rate and surface area of seed to be dried.
- The processes of evaporation and migration, occur simultaneously and continuously, first by removing the moisture from the evaporation surface and, due to potential difference, the moisture in seed center migrates to periphery, again being removed by evaporation, following the cycle, so internal water concentration gradient to seed surface, which disseminates and redistributes seed moisture content.
- Evaporation occurs due to water vapor partial pressure gradient between seed surface and drying air.
- The drying rate is ratio between mass amount of water withdrawn over time which depends on initial product moisture content and air flow, drying method, type of dryer, product characteristics, quantity and depth of product mass for air to cross.
- The drying rate referred to as drying speed is important to avoid seed deterioration by drying delay, but such velocity should be such that it does not create excessive stresses on seeds by heat or moisture gradient.

4.5.3 Dryers used in artificial drying

- The dryer forced the air through the seed bed and air is heated using flat plate instead of conventional means.
- In modern automated storage system, drying takes place at one of the two points either in prestorage dryers or in store dryer.
- In prestorage dryers, ambient air is used in continuous flow dryers, and heat is generated by thermostatically controlled furnace which is powered by electricity, diesel, or gas where heat is supplied by direct or indirect way.

- Indirect way is preferred due to separate outlet for combustion product not through seeds.
- In batch dryers, seeds are fed in properly defined batches, whereas in continuous flow dryers, the grains are flowed into the system and recovered at desired moisture content.

For artificial drying of grains, two types of dryer are used :

Static or discontinuous dryers or batch dryers

- Such dryers are less expensive and can handle medium quantities of grain.
- Floor dyers are good for all seeds of various sizes.
- Bag dryers are good for small seed grain lots to be used as seed where many varieties are handled where all types of grains are in sacks.
- Box dryers are good for large seed grain lot where slow drying is required.

Continuous dryers

- These are high-flow dryers that need a more complex infrastructure, complementary equipment and special planning.
- They are appropriate for big centers, silos or warehouses, where very large quantities of product are handled.
- Continuously flowing vertical dryers: Good for quick and normal drying and free flowing grains of single variety which can handle large volume of grain and are not good for small lots of different varieties.
- Continuously flowing belt dryers: Good for quick and normal drying seeds, small to very small lots and chaffy seeds.
- Rotary dryers: Good for quick and normal drying grains, chaffy grains with moderate moisture which are normally for large capacity drying and not for grains with high moisture.

4.6 Taking random samples of seeds for testing moisture level after drying them for time specific to different variety of crops as mentioned in the SOP

4.6.1 Seed Moisture Content

- Small change in seed moisture content has large effect on storage life of seeds, so it is important to know moisture content so as to make reasonable accurate prediction of possible storage life of each accession.
- The moisture content of seed sample is loss in weight when it is dried.
- Moisture content is expressed as percentage of weight of original sample.
- It is one of the most important factors in maintenance of seed quality.

- The seed moisture content can either be accurately determined experimentally by scientific techniques, or can be predicted from the given information.
- Moisture contents will differ from seed to seed in same accession, but should not be sufficiently different to make more than about 1% difference in moisture contents between seeds of same accession and different accessions of same species.

4.6.2 Determining seed equilibrium moisture content

- Take the required seed samples after drying process
- Take two lots of seeds of species of known weight and leave them in open containers in an environment
- Measure relative humidity of atmosphere where seeds have been held using an aspirated hygrometer.
- Measure the mean temperature.
- Remove sample of seeds and do accurate determination of moisture content for each seed lot.
- The mean of two tests can be used as guide since other seed lots of same species needs to be equilibrate at similar moisture content under certain conditions.
- Repeat determination the following day to make sure that moisture content is stable and that equilibrium has been reached,
- Write details of equilibrium moisture contents at different temperatures and relative humidities.
- Repeat the process, if required moisture content level is not achieved.

4.7 Stop drying seeds and send them for further processing once moisture level comes down to optimum as mentioned in the SOP

4.7.1 Checking seeds after drying

- Seeds which were harvested dry have enough moisture content when they are dried under sun for 2-3 days.
- If seeds were harvested wet or washed before drying, in such case 3-5 days of sun drying is enough after they have been air-dried for 1-2 days.
- Seeds have distinct sounds when their moisture content is already low while large, thin seeds will break with snapping sound when twisted between the fingers.
- Large, thick seeds will break with cracking sound when bitten between the front teeth.

- Small seeds will break with cracking sound when squeezed between the fingernails.
- Seeds have distinct tinkle when they are well-dried.

4.7.2 Seed processing after removing moisture

- After seeds have been cleaned and graded, they are ready for sowing in nursery.
- If seeds are to be kept in storage, it is necessary to check their moisture content and if necessary, adjust it to optimum level for storage.
- If the moisture content of seeds after checking is found optimum, then such seeds can be send for either sowing in fields or can be send for storage.

5

Achieve Productivity and Quality Standards

5.1 Ensure selection of healthy and viable seeds for processing

5.1.1 Introduction

- Seed is a fundamental requirement to grow most of the crops.
- It is that part of plant which is used for propagation, plantation or regeneration purpose.
- Vegetable seeds are costly and their wastage during sowing or handling increases the cost of cultivation.
- Healthy and good quality seeds lead to healthy crop, so selection of seeds is important.

5.1.2 Seed selection

Seed processing worker should ensure that the seeds selected for further processing should:

- be whole seed and not broken or crushed seeds.
- have high sowing quality.
- have high germinating capacity.
- be free from infection.
- not be mixed with seeds of weeds or with other seeds.
- be disease-resistant.
- be purchased from good shops or supplied by good seed agencies (certified seeds).

5.1.3 Examine seeds before processing

For proper processing of seeds, seed processing worker should follow the steps:

STEP 1: Examine the state of seeds supplied by:

- treating each accession separately

- emptying contents of seed packet in labelled tray.
- checking whether seeds are in dry fruits or fleshy fruits or have already been cleaned.

STEP 2: Clean seeds having debris by:

- checking seed dryness as moist seeds can be damaged during cleaning.
- placing moist seeds in drying room or in shady place to dry.
- predicting seed moisture content between 12%-16% for safe seeds cleaning.
- cleaning seeds from fruits using hands.
- small threshing machines and prevent mixing of seeds from different accessions.
- using gradated sieves.
- winnowing, blowing gently on seeds or using seed blower.

STEP 3: Examining cleaned seeds for insect and fungal damage

- If seeds are infested with insects or infected with fungi, isolate seeds and dry infested seeds in sealed containers with silica gel to prevent spreading of insects.
- Drying seeds to low moisture content will prevent further growth of fungi and development of further reproductive cycles of insects, but it will not kill them.
- Storage at sub-zero temperatures will kill adult insects but not fungi.
- If seeds are to be stored above 0°C, place in freezer for 7 days to kill the insects before continuing with normal packaging and storage procedures.

STEP 4: Examine cleaned seeds for physical damage

- Spread the seeds on flat well lighted surface.
- Examine the seeds carefully for physical damage and empty the seeds.
- Discard any visually damaged, shrivelled or empty seeds.
- Empty seeds may be difficult to detect unless seeds are dissected or tested with special X-ray equipment to determine percentage of empty seeds.
- Empty seeds are lighter than full ones and can be separated by using seed blower.

5.2 Ensure proper pre-cleaning of seeds by removing appendages and particles such as trash, stone clods and dust

For proper processing of seeds, seed processing worker should ensure that the raw seeds should be cleaned well and should not contain any:

- appendages

- large pieces of trash
- stones
- debris
- dust

5.3 Clean seeds in a way that causes least damage to seeds and does not waste good seeds

- Seed cleaning normally starts with pre-cleaning procedure where large material like leaves, twigs and entire empty fruits are removed which is done in connection with extraction and is essential for subsequent cleaning where machines are employed.
- Cleaning large quantities of seeds is more efficiently done by mechanical equipment, especially where combined machines can be used for extraction and cleaning.
- Different kinds of seeds can be separated when they differ in one or more physical characteristics such as size, shape, length, weight, colour, surface texture, affinity to liquids, electrical conductivity, etc.

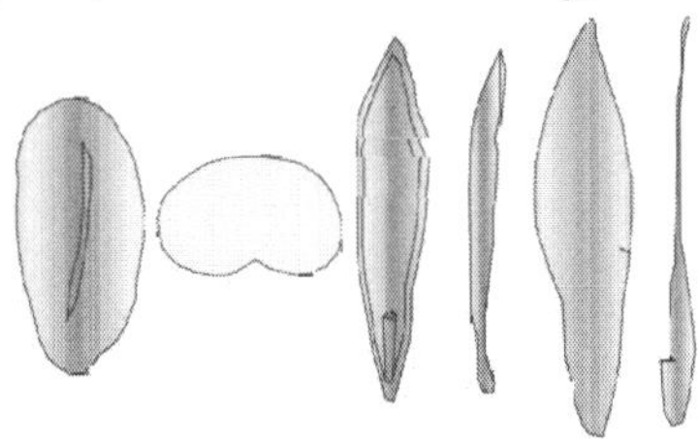

Seed worker should note that seed cleaning should be done in such a way that it should cause least damage to seeds and does not waste good seeds by following:

- If clean seed fraction still contains debris after first cleaning, then cleaning machine should be adjusted and cleaning should be done repeatedly.
- Cleaning of seed should be done as per type of seed as hard seed can be cleaned using machine or by washing under water or can be run through a screen.
- Dry seed can be crushed by hand and rubbed over a screen to separate the seed from the chaff.
- Seed worker can also blow on seed to separate lighter chaff from heavier seed.
- Seeds which are very soft can be macerated by hand or can be kept in water to float.
- Larger seeds can simply be separated from their husk, pod or cap by hand.

5.4 Use appropriate seed cleaning equipments like seed blower, graded shieves, scalpers, debearders and huller-scarifier etc

5.4.1 Seed blower

- Seed blower makes separation of undesired materials accurately by controlling upward air flow by transparent plastic column for purity testing, docking and cleaning of grain/seed samples.

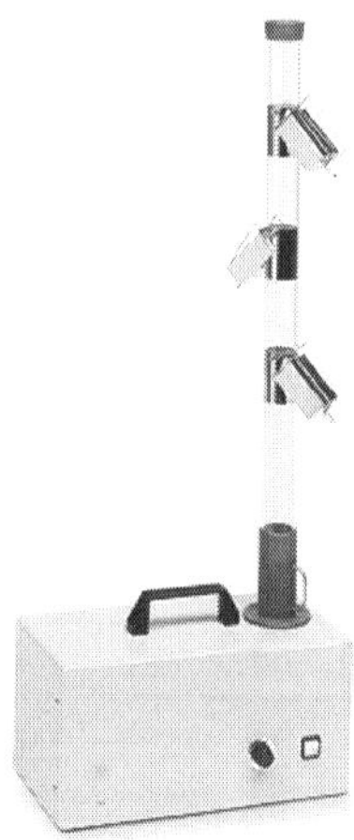

- It serves as quick and easy separation of light and heavy seed.
- In seed blower, air flow can be controlled to get most difficult sensitive air separations.
- In this, seeds are filled in drawer of glass cylinder and fine-meshed sieve closes the upper end of cylinder.
- A speed-controlled blower in cylinder, generates constant upward air column.
- As per the size and weight, seed will fall in one of collecting vessels being installed at cylinder where light seed gets deposited in upper collecting vessel while heavy seeds will be deposited in lower collecting vessel.

- In this, sieve at the bottom of cylinder avoids penetration of seed in blower.
- Once blower is disconnected, sieve contains deposit of residues and of small stones.
- The collecting vessels are equipped with snap buckle clip which allows easy removal.
- Spread seeds on a flat well-lit surface of contrasting colour such as an illuminated table to remove damaged seeds, seeds of different species etc. if any

5.4.2 Graded Sieve

Sieves are made of flat perforated plate or wire mesh which are three dimensional like funnel shaped. Following are the graded sieve machines used for removing dust, debris, trash etc from seeds which includes:

Machine 1:

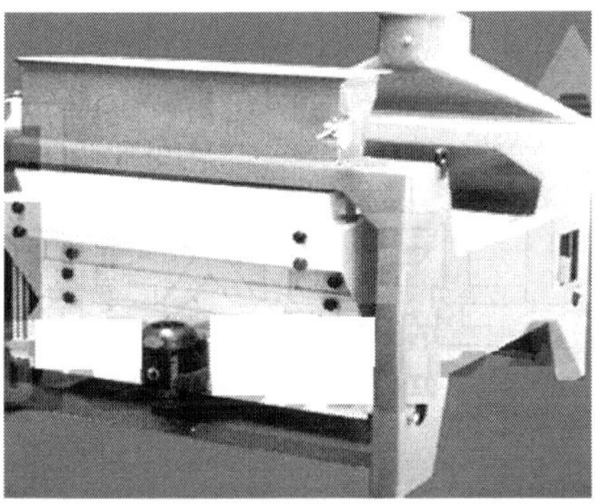

- The graded sieve machine is used for cleaning/ grading of oilseeds using set of two sieves, to remove coarse and fine impurities and attachment to aspiration system which is used for removing low-density matter.
- The machine comprises of frame structure which is made of steel channel with fully enclosed sieve box that is suspended at its four corners on rubber bushes.
- In this, box has two decks with ball cleaned sieves and vibrator drive is located in center.
- The vibratory movement of sieve increases the capacity on a given sieve-surface and sieve effect.
- The capacity of the machine ranges from 1-10 TPH
- The machine is used for thorough removal of dust, sand, stones or unwanted particles and is easy to operate.

Machine 2:

- ❖ The graded sieve machine is suitable for cleaning of all types of grains and seeds.
- ❖ It is used to separate out impurities, which are bigger/smaller than seed size.
- ❖ The capacity of the machine ranges from 1-3 TPH
- ❖ The machine is used for thorough removal of dust, sand, stones or unwanted particles and is easy to operate
- ❖ The machine enhances the life of spares of oil expellers where magnet placed at discharge end is used for removal of iron pieces

Machine 3:

- ❖ The graded sieve machine is a type of rotary seed cleaner which is suitable for cleaning of all types of grains and seeds.
- ❖ It is used to separate out impurities, which are bigger/smaller than seed size.
- ❖ The capacity of the machine ranges from 1-3 TPH
- ❖ The machine is used for thorough removal of dust, sand, stones or unwanted particles and is easy to operate

5.4.3 Scalpers

- Scalper is an equipment which is used when seed initially comes from the field.
- Scalpers can be installed ahead of main receiving elevator in cleaning plants, dryers, or storage facilities.
- Scalping is a rough cleaning operation which is normally done at high capacity to screens off foreign material which are bigger than crop seed.
- The scalped material contains stems, green leaves, weed seeds, insects, and trash, which shows half or more of total lot.
- The removal of large bulk of waste early will help in reducing later handling and storing needs and improves cleaning efficiencies.
- Removal of trash and high-moisture green material will lower drying costs and allow safe storage.
- There are many types and sizes of scalpers as one with reel of perforated metal screen that turns central shaft and is inclined from horizontal.
- In this, the material that feeds in higher end gets tumbles in reel till crop seeds drop by perforations while large trash stays in reel and gets discharged separately.
- The other form of scalper makes the same kind of separation with single, flat, perforated screen that is mechanically shaken.

5.4.4 Debearders

- Many seed lots gets cleaned directly after scalping, while others require further conditioning.
- In many cases, lot has unthreshed seed in form of pods, heads, clusters or doubles while some seeds like barley have awns or beards.
- To minimize seed loss, unthreshed units needs to be broken into single seeds and awns should be removed for proper separation.
- The idea of such machines is to do certain functions which completes threshing process.
- The debearder has horizontal beater assembly which rotates inside a steel drum.
- In this, beater is made of shaft with projecting arms that are pitched to move the seed mass through the drum.
- In this, stationary posts will be adjusted for clearance with arms, that is protrude inward from the drum which prevents mass from rotating with beater.

- While operating such machine, vigorous rubbing of seeds, pods, heads and doubles are created against the arms, posts and each other.
- In this, time that seeds remain in unit varies by regulating a weighted discharge gate.
- Here severity of action is controlled by exposure time, beater speed and clearance among beater arms and posts.
- The machine is used extensively to remove awns and hairs from certain grains, native grasses, and other seeds.
- In this, seeds with appendages are difficult to handle, clean and plant as they tend to interlock and cause undesirable clustering.
- A properly adjusted machine can greatly improve flow properties of difficult-to-handle seed.
- The machine involves in performing debearding of barley, clipping seed oats, threshing whitecaps in wheat, defringing carrot seed, decorticating sugar beet seed, subdividing fescue doubles, and removing fuzz from bluegrass.

5.4.5 Huller-scarifier

- Once the scalping and debearding of seeds are done, many kinds of seeds get cleaned without further conditioning, but many legumes and grasses needs hulling, scarifying, or both.
- Hulling is removal of outer coat or husk to improve seed's cleaning characteristics or its planting qualities.
- When the husk of seeds gets removed, the seed is readily handled in cleaning and planting equipment, and husks are impermeable to water and thus prevent termination.
- In scarification, seed coat is scratched or ruptured, as scarifying is done with hard' seeds which will not absorb water and germinate promptly.
- Hullers and scarifiers usually abrade the seeds between two rubber-faced surfaces or impel seeds against roughened surfaces like sandpaper or carborundum stone.
- In this, severity of abrasion or impact needs to be controlled accurately to prevent seed damage.
- Seeds of high moisture content are harder to hull or scarify than seeds with less moisture while clean seed scarifies more uniformly and with less damage than dirty seed.
- Many seed loses viability quickly after being hulled or scarified, so such processes should be delayed for such seed until shortly before planting time.

- Other seeds that maintain viability well after being hulled and scarified can be processed soon after harvest and stored safely until the following season.
- Hulling and scarifying are done separately or jointly as per presence of unhulled seed, hard seed, or both.
- Some seeds which may needs hulling are bermudagrass, bahiagrass, buffalograss, lespedeza while some seeds that needs scarification are alfalfa, subclover, white clover, suckling clover, hairy indigo, crotalaria, and wild winter peas.
- Seeds which may need hulling and scarification are sweetclover, sourclover, sericea lespedeza, crownvetch, and black medic.

5.5. Use appropriate seed drying equipments like Dehumidifier, seed-drying cabinets, dried silica gel etc. and methods for drying the seeds to assure quality

5.5.1 Dehumidifier

- Seeds can be preserved for long period by lowering the moisture content of seeds to around 5% by drying the seeds.
- Conventionally, seeds are dried at elevated temperatures which destroy their germination potential.
- Bry-Air Seed Dryers, based on the state-of-the-art Desiccant Dehumidification Drying Technology, dry the seeds to the required level in approximately eight hours without affecting their quality.
- Dehumidifiers offered are designed to efficiently remove excess moisture from atmosphere of laboratory, work place or from storage facility.
- Dehumidifiers come in two primary varieties where thermal condensers are common, and include standard air conditioners while desiccant or absorptive dehumidifiers are an emerging technology which can provide relief in settings where thermal condensers aren't as effective.
- Dehumidifier uses refrigeration principles same as the way an air conditioner does, but it doesn't cool the air, but it warms the air slightly.
- In this, warm humid air from structure/place is drawn in cold dehumidification coil of dehumidifier where temperature of air is low below the dew point condition making water to get collected on dehumidification coil.

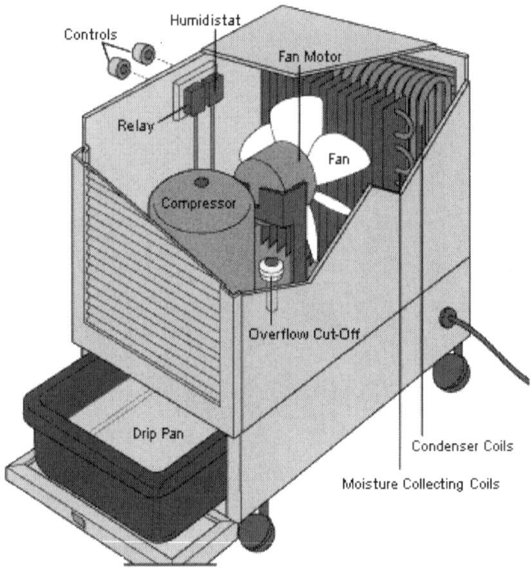

A dehumidifier is a box that contains

- ❖ cold evaporator coils,
- ❖ warm condenser coils,
- ❖ a fan, and
- ❖ a reservoir or drain to catch water

The fan blows moisture-laden room air over the cold coils, where water condenses and drips into the pan or drain. The drier air then blows across the warm coils and back into the room.

Although a dehumidifier does not cool the air, it makes it more comfortable because it decreases the humidity. But a dehumidifier does not function well under about 65 degrees F.

A dehumidifier's fan must have electric power, the evaporator coils must contain a proper charge of refrigerant, and the controls must work for the dehumidifier to operate.

In the same way that a thermostat controls a heater or air conditioner, a dehumidifier is cycled off and on by a humidistat that measures a room's humidity.

Application of Dehumidifier

Some commercial application of dehumidifier are:

- ❖ It lowers the relative humidity to facilitate manufacturing and handling of hygroscopic materials

- ❖ It lowers the dew point to prevent condensation on products manufactured in low-temperature processes
- ❖ It provides protective atmospheres for the heat treatment of metals
- ❖ It controls humidity in warehouses and caves used for storage
- ❖ It preservers ships, aircraft, and industrial equipment that would otherwise deteriorate.
- ❖ It maintains a dry atmosphere in a closed space or container, such as the cargo hold of a ship or numerous static applications
- ❖ It eliminates condensation and subsequent corrosion
- ❖ Drying air to speed the drying of heat-sensitive products, such as candy, seeds, and photographic film.
- ❖ Drying natural gas.
- ❖ Drying gases that are to be liquefied.
- ❖ Drying instrument air and plant air .
- ❖ Drying process and industrial gases .
- ❖ Dehydration of liquids .
- ❖ Frost-free cooling for low-temperature process areas such as brewery fermenting, aging, filtering, and storage cellars; blast freezers; and refrigerated warehouses.
- ❖ Frost-free dehumidification for processes that require air at a subfreezing dew point humidity.

5.5.2 Seed-drying cabinets

- ❖ Drying cabinet is basically of two-door which has built-in cooling unit and is attached to a dehumidification unit.
- ❖ In this, moist air is removed by connecting the outlet to seed bank's duct system.

- In such cabinet, seed collections of species gets moisture status below 20% eRH in few weeks, when set at 15% RH and 15°C.
- The drying time of seeds in this cabinet depends on seed size and collection volume.
- The advantage of drying cabinet system is that it is relatively cheap, easy to maintain and can easily be moved around within seed bank.
- In this, constant temperature and humidity is generated throughout.
- The drying seed cabinet is easy to clean and insects and other pests can be kept under control, but as all cleaning, counting and testing of seeds needs to be done outside the cabinet, the seeds may become partially rehydrated during processing.
- In peak collecting times, such cabinet becomes crowded and collection will be difficult.

5.5.3 Dried silica gel

- Silica gel is present as clear beads or indicating beads that changes colour as per moisture status.
- Silica gel is useful desiccant which can be used for drying seeds.
- The gel is available as either non-indicating clear beads or indicating beads that incorporate non toxic indicator which changes colour as per state of hydration.
- In this, orange/green indicator is orange when dry with RH below 20% and dark green, when RH is above 50%.
- The colour change between two extremes which takes place at 25% RH close to ideal 15% RH that is used for long-term seed storage.
- A mix of 10% indicating to non-indicating beads are routinely used in MSB.

For drying the seeds, dry the seeds at ambient conditions before starting and follow the steps:

- Fill the container to around 20% by volume with oven-dried silica gel beads while mix of 10% indicates recommendation of non-indicating beads.

- Keep seed collections in cloth or paper bags in container and ensures that proper air gets circulated.
- Now maintain a minimum weight ratio of 1:1 silica gel to seed material.
- Put the drying container out of direct sunlight, in cool place.
- On weekly basis, mix the silica gel in bottom of container and measure its eRH by hygrometer.
- Maintain the silica gel at less than 15% RH.
- If hygrometer is not available, make weekly checks of silica gel in bottom of container by looking at colour of indicator.
- When silica gel absorbs moisture from the air in container, in such case, indicator will change from orange to green.
- Mix the seeds in each collection and measure the eRH.
- Seed moisture status can be assessed by including 1 gram indicating silica gel sachet in each collection and compare it with colour chart.

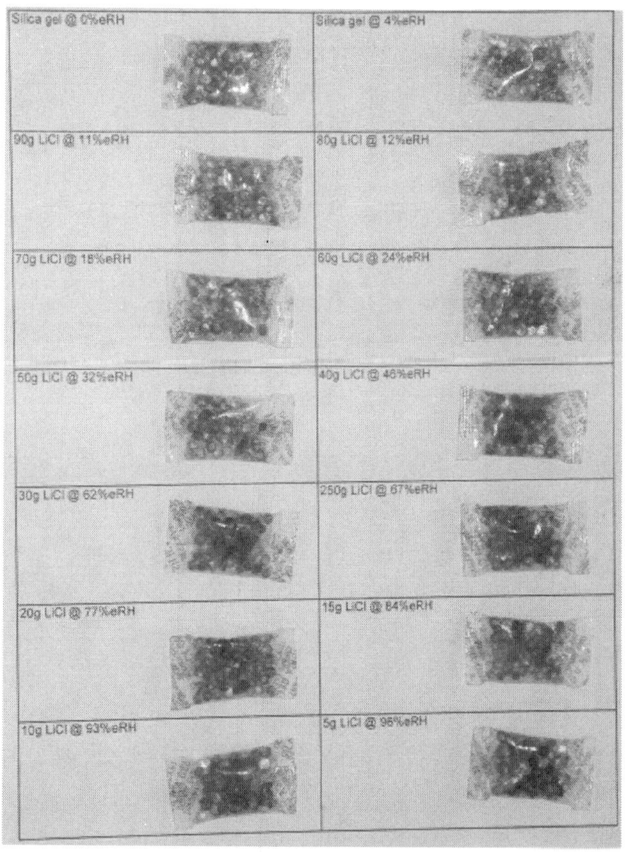

Fig Silica Gel sachets at various LiCl concentration

- ❖ The seeds takes at least one month to dry which depends on initial moisture status of seeds, collection volume, type of seeds and moisture status of silica gel.
- ❖ Small seeds will dry more quickly so it needs more monitoring regularly.

5.6 Ensure that all seed must dry uniformly

- ❖ Drying of seeds involves in removing of moisture from the seeds to maintain its viability and vigour during storage, which may otherwise deteriorate fast owing to mold growth, heating and enhanced microbial activity.
- ❖ Seed drying allows early harvesting, long term storage of seeds, more efficient use of land and manpower, use of plant stalks as green fodder and production of high quality seed.
- ❖ For uniform drying of seeds, seed needs to be spread in thin layer under heavy sun and should be stirred frequently.
- ❖ Once the above process is completed, seeds are collected and are kept.
- ❖ Seed processing worker can manually check the dryness of seeds as dry seeds has 10% moisture content which is good for seed processing and storage.

5.7 Prevent mixing of seed from different accessions during cleaning and drying of seeds

- ❖ Once the seeds are threshed from straw either by impact, friction or combing action manually or by threshing machines, they needs cleaning.
- ❖ Seed cleaning is the removal of debris, physical contamination, inert material, damaged and infected seeds and seeds of other species.
- ❖ Small seed cleaning machines can be used for cleaning of certain seeds.
- ❖ The milled seeds contain dried plant parts which gets cleaned in the machine.
- ❖ Different types of seeds with other parts are cleaned in such machine, so once the first seed lot gets cleaned, seed worker/machine operator should

thoroughly clean the machine, so that seeds of new seed lot will not get mixed with previous seed lot during cleaning.

- Seed processing worker needs to check that machine operator should thoroughly clean the machine, before cleaning next seed lot.
- Once the seeds of different accession are cleaned, they should be kept in different storage containers.

Unit 2: Treat Seeds with Chemicals

1

Prepare Chemical(s); Tools and Equipments for Application

1.1 Receive required chemicals like pesticides, insecticide and fungicide from supervisor as per types of crop and varieties of seed

1.1.1 Introduction

- To protect the seeds from seed-borne or soil-borne pathogenic organisms and storage insects, seed needs treatment by use of chemicals such as fungicide, insecticide or combination of both.
- Seed processing worker ask from store supervisor the type of chemicals to be used for treating seeds.
- Seed treatment is important in protecting seeds and seedlings from seed and soil borne diseases and insect pests that affects crop emergence and its growth.
- Treating seeds with chemicals reduce potential risks to workers, minimize potential runoff to waterways, and lower the overall amount of pesticide applied in the environment.
- Different chemicals such as fungicides, insecticides, inoculants, plant growth regulators, fertilizers and fertilizer are used in treatment of seeds which are used alone or in combination to address or prevent number of pests, diseases and nutrient deficiencies so as to enhance the plant growth.

1.1.2 Seed Treatments

- Seed treatment describes products and processes where usage of particular products and specific techniques improve the growth environment for seed, seedlings and young plants.
- Seed treatment ranges from basic dressing to coating and pelleting.
- Seed Dressing is a common method of seed treatment where seed is dressed with dry formulation or wet treated with slurry or liquid formulation. In this, dressings can be applied at farm and industries. In this, low cost earthen

pots can be used for mixing pesticides with seed or seed can be spread on polythene sheet and required quantity of chemical can be sprinkled on seed lot and mixed mechanically by processing workers.

- Seed Coating involves special binder with formulation to enhance adherence to seed. In this, coating needs advance treatment technology by industry.
- Seed Pelleting is sophisticated seed treatment technology which results in changing physical shape of seed to upgrade palatability and handling. In this, pelleting needs special application machinery and techniques and is most expensive.

1.1.3 Knowledge about chemicals

Seed processing worker should have an idea about basic chemicals that can be used for treating seeds against pest, insects which should include:

- Herbicides are substances that kill or inhibit growth of unwanted plants (weeds)
- Insecticides are substances that kill arthropod pests, i.e. insects and mites
- Fungicides are substances that destroy or prevent the growth of pathogenic fungi

1. Fungicides and insecticide:

- Fungicides and insecticide treatment will help crop growers to protect seeds and seedlings from disease and to fight pests that strike early in season when seedlings are most vulnerable.
- Fungicide applies to synthetic chemical compounds which kills fungi or inhibit their growth, but there are certain biological organisms that can also be used to control fungal infections such as mildews, rusts and leaf spots.
- Fungicides prevent and cure diseases which can have adverse effects on crop yields and quality.
- Treating seed with fungicide will inhibit surface-borne fungi on seed coat and protects seed against soil-borne disease.
- Many fungicides offer protection against development of fungi, and others cure the plant by eliminating the fungus.
- All modern fungicides degrade in aerobic soils which are environmentally friendly where rate of degradation depends as per temperature and humidity.
- Since fungi rapidly develop resistance to chemicals, it is necessary to be able to draw on fungicides from different chemical classes and with different sites of action.

2. Herbicides

- Herbicides are used to control the growth of unwanted plants where modern herbicides act by restricting growth by inhibiting the action of one or more of receptors that catalyze reactions which are essential to growth of plant.
- Auxins kill by overstimulating growth, so with selective herbicides, either the target in weed is affected more than that of crop where herbicide degrades more quickly in crop, or uptake or translocation of active ingredient that differs from weeds.

Types of herbicides

Translocated herbicides:

- They move to the site of action by transport mechanisms in plant; xylem and phloem.
- The xylem transports water and nutrients from the soil to growth sites and phloem transports products of photosynthesis (for instance, sugars) to growth and storage sites.
- It may take up to two weeks for symptoms to develop on the target weeds depending on herbicide rate, conditions and species.

Contact herbicides

- They have limited movement within the plant, so complete coverage of target is critical.
- Compared to translocated herbicides (glyphosate), contact herbicides (paraquat, oxyfluorfen, diquat and bromoxynil) tend to show symptoms rapidly, usually within 24 hours.

Selective herbicides

- It kill target weeds and not desired plants (crop or pasture) when applied at specified application rate.

Non-selective herbicides

- It is also called as knockdown herbicides like glyphosate or paraquat that damage most plants.

Residual herbicides

- It remains active in the soil for an extended period of time and acts on successive weed germinations.

Non-residual herbicides

- It includes non-selective paraquat and glyphosate.
- It has little or no soil activity and are quickly deactivated in the soil.

- ❖ They are broken down or bound to soil particles and is less available to growing plants.
- ❖ They have little or no ability to get absorbed by roots.

Post-emergent and pre-emergent

- ❖ These are target and timing of herbicide application.
- ❖ Post-emergent is foliar application of the herbicide after the target weeds have emerged from the soil.
- ❖ Pre-emergent is the application of herbicide to the soil before weeds have emerged.
- ❖ Herbicide mixtures and sequential applications
- ❖ It involves application of more than one herbicide to increase spectrum of weed species controlled for resistance management.
- ❖ A mixture involves the application of multiple products in a single application.
- ❖ Where herbicides are antagonistic and cannot be mixed together in a single tank, they are applied sequentially.

Chemical types

Tata Mida 70 WS

- ❖ Its Technical is Imidaclorpid 70 WS.
- ❖ It is a novel insecticide belonging to new generation NEONICOTINOID group of chemicals.
- ❖ It is an excellent insecticide for control of sucking pests in Cotton, Okra and Chilli.

Glazer 35 WS

- ❖ Its technical is Metalaxyl 35 WS.
- ❖ It is used for protection from pre and post emergent seedling blights and fungal seed rots in Maize, Bajra, Sorghum & Sunflower.

Tata Mida 600 FS

- ❖ Its technical content is Imidaclorpid 48% FS.
- ❖ It is a new generation insecticide belonging to NEONICOTINOID group.
- ❖ It is less toxic, systemic insecticides with a broad spectrum of activity for controlling sucking pests such as Jassids, Aphids, Whitefly and Thrips in cotton, okra & Sunflower.

Captaf 50 WP

- Captaf 50 WP is an organic, broad-spectrum, contact fungicide based on CAPTAN as the active ingredient.
- It is ideally suited to control diseases like Scab, Brown rot, Damping off, Downy Mildew, Fruit rot, Clump rot, various Leaf -spots including Tikka disease, Early and Late Blights and Dry root rot diseases effectively.

Trooper

- Trooper is a broad spectrum fungicide for the control of seed and soil borne diseases.
- It has both systemic and contact mode of action.
- It also acts as a plant growth stimulant.

COMMON NAME	CHEMICAL NAME	REGISTERED TRADE NAMES
thiram	tetramethylthiuram disulfide	Arasan 50, Arasan 42S, Vancide TM, Ever-shield T
maneb	manganous ethylenebisdithiocarbamate	Dithane M-22, Manzate D
maneozeb	manganous ethylenebisdithiocarbamate —coordinated with zinc ion	Dithane M-45, Dithane Z-78
captan	N-[trichloromethylthio]-4-cyclohexene-1,2-dicarboximide	Captan SP, Orthocide 50, Evershield CM
folcid	tetrachloroethylmercapto-cyclohexene-dicarboximide	Difolatan 4F
dichlone	2,3-dichloro-1,4-naphoquinone	Phygon XL
nabam	disodium ethylenebisdithiocarbamate	none
bordeaux	copper sulfate + calcium hydroxide 5:5	none
mercury	methyl-mercury proprionate	Metasol MP
none	dimethylaminobenzenediazo sodium sulfonate (p[dimethylamino] benzene-diazo sodium sulfonate)	Dexon
none	pentachloronitrobenzene + 5-ethoxy-3-trichloromethyl 1,2,4 thiadiazole	Terraclor Super X, Terracoat L-21
benomyl	(1-[butylcarbamoyl]-2-benzimidazole carbamic acid)	Benlate
copper	copper hydroxide	Kocide 101
thiabendazole	2-[4′-thiazoyl] benzimidazole	TBZ
chlorothalonil	tetrachloroisophthalonitrile	Daconil 2787
dithianon	1,4-dithiaanthraquinone-2,3-dicarbonitrile	Delan
none	(a copper zinc chromate complex)	Miller 658
TCMTB	2-[thiocyanomethylthio] benzothiazole	Busan 72
metiram	zinc polyethylenethiuram disulfide	Polyram
chlorneb	1,4-dichloro-2,5-dimethoxybenzene	Demosan 65W

1.2 Keep different type and form of chemicals as per SOP in safe manner at designated place

1.2.1 Introduction

Pesticides are toxic to pests and humans, but are not hazardous to humans and non-target animal species if suitable precautions are taken. Many pesticides causes adverse effects if intentionally or accidentally ingested or are in contact with skin for longer time.

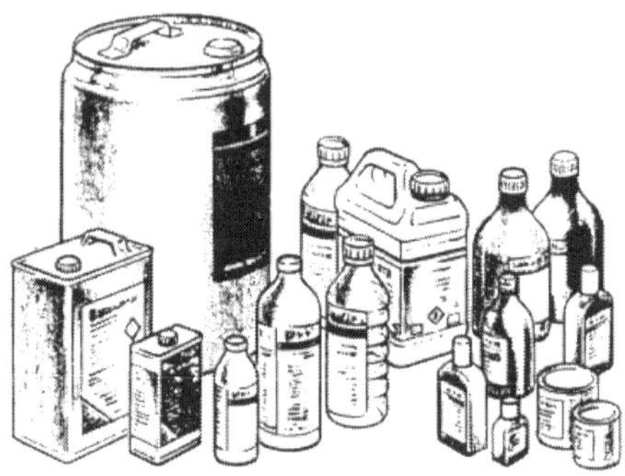

Pesticide particles may be inhaled with the air while they are being sprayed. Also, the risk is the contamination of drinking-water, food or soil. Special precautions must be taken during transport, storage and handling.

1.2.2 Before storing pesticide

- Read the label to learn how to store the pesticide correctly
- Pesticides should never be stored with food or feed
- Many pesticides should not be stored where they may become contaminated by other pesticides
- Many pesticides have heat and cold recommendations, or should not be exposed to UV radiation
- The label will tell you if there are any special storage precautions to follow.

1.2.3 Role of pesticides and fertilizer applicator to store pesticides

Pesticides and fertilizer applicator should try to use all the pesticide in application tank. If pesticides remain, use them on other target locations. After emptying the tank, clean and store the equipment. The following storage criteria should be followed for pesticides labeled with signal words DANGER, POISON, or WARNING, or the skull and crossbones symbol. These procedures and criteria are not necessary for the storage of pesticides classed as less toxic or for those registered for use in the home or garden.

Site Storage

- Locate where flooding is unlikely.
- Locate where runoff will not contaminate any water system.

Storage Facility

- Dry, well ventilated, separate room, building, or covered area with fire protection (e.g., dry chemical fire extinguisher).
- Secured by fence and/or locked doors.
- Signs on rooms/buildings to provide hazard warning (e.g., DANGER, POISON, and PESTICIDE STORAGE).
- Movable pesticide equipment is labeled as contaminated and not removed from the site until decontaminated.
- Provision is available for the decontamination of personnel and equipment; contaminated water disposed of as excess pesticide; contaminated runoff collected and treated as excess pesticide.

Operational Procedures

- Store pesticide containers in rows with the labels plainly visible.
- Place contents from damaged containers in sound containers.
- If relevant, segregate pesticides by formulation.
- Store rigid containers in an upright position, with tight lids/bungs, off the ground, in a manner to permit access and inspection.
- Maintain a complete inventory indicating the number and identity of containers.
- Check containers regularly for corrosion and leaks.
- Keep suitable absorbent (e.g., vermiculite) on hand in case of spills.

1.3 Wear protective gears as per SOP

1.3.1 Introduction

Pesticides can pose hazards to humans as severity of harmful effect or poisoning depends on pesticide's chemical makeup and formulation into the body, amount that enters in body, and length of exposure. As per SOP, crop/seed worker while working with chemicals should follow SOP instruction and should wear personal protective gears/equipment which will greatly reduce the potential for dermal, inhalation, eye and oral exposure, and significantly reduces chances of pesticide poisoning.

Pesticide and fertilizer applicator should manage their risk by wearing protective clothing and equipment when handling or applying pesticides, so that risk of exposure should be avoided along with risk of pesticide poisoning. Understanding the toxicity of a product and the potential for personal exposure allows pesticide and fertilizer applicator to lower their risk which can also be concluded from risk formula as risk = toxicity x exposure. Toxicity of substance can't be changed, but risk can be managed, and the applicator is a manager. No matter how toxic a substance is, if the amount of exposure is managed, risk can be held to an acceptably low level.

Different pesticide products require different personal protective equipment. The type of personal protective equipment needed depends on toxicity of pesticide being used and its formulation. Some labels, mainly for agricultural pesticides are affected by Worker Protection Standard and specifically state that certain items of clothing, equipment, eyewear, footgear, and gloves must be used. Some of the PPE required are specific to early entry while others are specific to handling and applying, so the more toxic the pesticide is, greater will be the need for personal protective equipments.

Chemicals are mostly used for controlling disease, insects and weeds in crops which saves crop from pest attack only if applied on correct time. Such chemicals need to be applied on plants and soil in form of spray, dust or mist. These chemicals are costly, so equipment for uniform and effective application is essential. Farmers and seed growers are exposed to pesticides in many different situations such as in mixing, loading, spraying, storing, spillage and disposal

Pesticides enter in body through skin, breathing and mouth

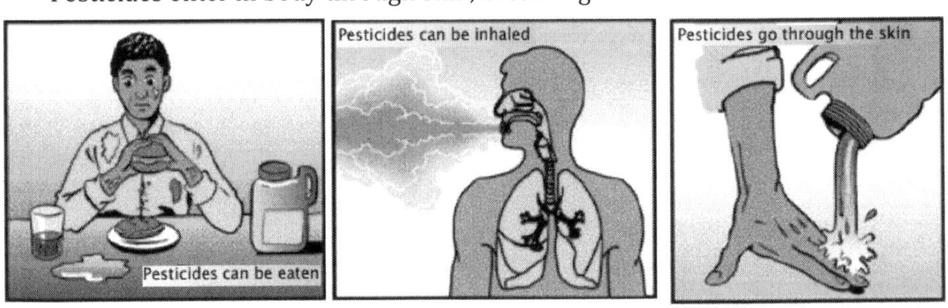

Standard operating procedure is intended to provide general guidance on how to safely work with highly reactive/unstable materials. The effects of chemical

exposure to farmers or farm workers while working on fields depend on the type of chemical and degree of exposure. If chemicals are swallowed, absorbed through the skin or inhaled as mist, vapour or dust, then there appears a different immediate effect such as poisoning, headache, nausea, vomiting, diarrhea, pinpoint pupils, dizziness, fine muscle twitching, increased bronchial and lacrimal secretions, skin rashes and irritation, chemical burns.

1.3.2 Selecting Right Personal Protective Equipment

- If a pesticide label does not have specific personal protective equipment requirements, pesticide and fertilizer applicator should always take reasonable precautions and use common sense.
- Use route-of-entry and specific action statements from the product label to determine the type and degree of protection needed to handle the pesticide safely.
- Liquid pesticides often are more hazardous to use than dry formulations, and extra protection is warranted while mixing and/or loading such pesticides.
- Recognize that in cases where there will be pro-longed exposure to spray or where the application is being made in an enclosed area, then pesticide and fertilizer applicator should use extra protection.

1.3.3 Use of Protective Clothing

- Whenever pesticide and fertilizer applicator uses pesticides, they should atleast wear a long-sleeved shirt, long pants, shoes, socks, and chemical resistant gloves.

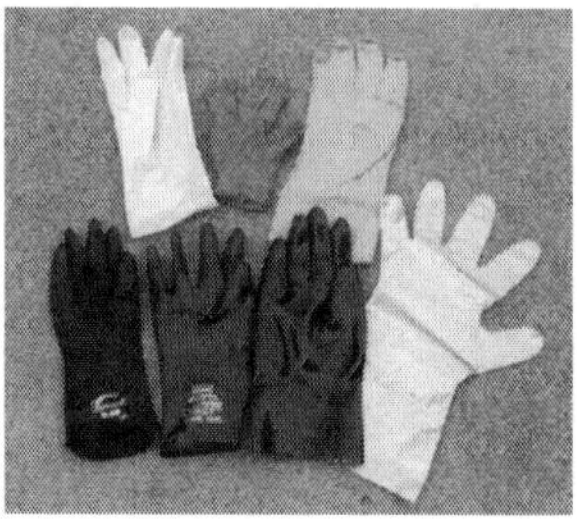

- Many labels will require pesticide and fertilizer applicator to wear more than this, as per product toxicity and usage.
- To reduce pesticide penetration, pesticide and fertilizer applicator should select garments made of tightly woven fabrics.
- Disposable coveralls made of Tyvek, provide adequate protection to pesticide applicator under most conditions.
- Protective suits made or coated with butyl rubber, neoprene, PVC, or of polyethylene fabrics are required for certain applications.

- Shoes and socks should be worn to avoid flip-flops and exposure to feet.
- Wear chemical resistant boots while working with highly toxic liquid pesticides and on prolonged exposure to pesticide spray.
- Pesticide and fertilizer applicators who mix and load liquid concentrates should wear chemical resistant aprons.

1.3.4 Protecting Head, Eyes, and Hands

- Protection for head is advisable and in some cases is specifically required.
- A wide-brimmed, easily cleaned hat will keep pesticides away from the neck, eyes, mouth and face is required.
- Avoid hats with cloth or leather sweatbands as these will absorb pesticides.

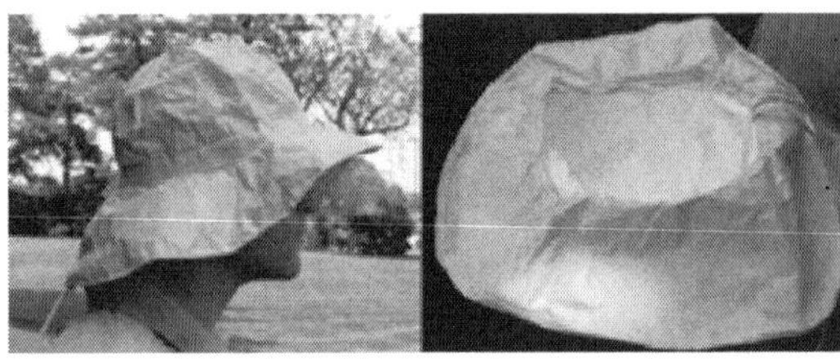

- Labels that specify the use of headgear are found on highly toxic liquid concentrates.
- When working with such pesticides, wear a chemical resistant hood or plastic hard hat with plastic sweatband and rain-trough edge to keep drips off the neck and back.
- Pesticides are readily absorbed through eyes and can cause eye injury.
- When the labels for liquid pesticides include precautionary statements, it generally indicates the need for eye protection.
- Use goggles or safety glasses when the label requires it.

- Some goggles have wider bridge over the nose to be compatible with respirators.
- Goggles will provide adequate protection if they have the right type of venting.

1.3.5 Protecting Lungs

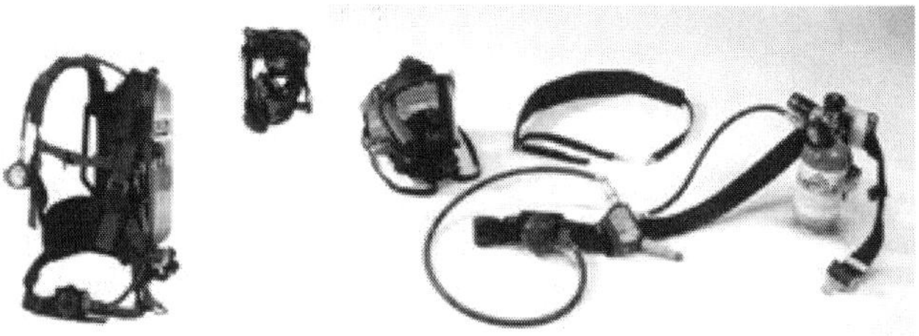

- Lungs and lining of respiratory system readily absorbs pesticide dusts and vapors from the air.
- Respiratory protection is required whenever the label calls for it and is recommended during mixing and loading, even if not required by label.
- Respiratory protection is recommended whenever pesticide and fertilizer applicator will be exposed to intensive concentrations of pesticide dusts, fumes, or vapors.
- The type of respirator which pesticide and fertilizer applicator uses, will depend on type and toxicity of pesticide, application site and other factors.
- Particulate respirators are acceptable when applying pesticide dusts and granules, and for protection against large droplets suspended in air.
- Many air-purifying respirators has tight-fitting mask with disposable cartridges or canisters whose design is of half-mask or full-face.
- Air-purifying respirator with suitable cartridges/canisters facilities are required for protection against vapours.
- Such respirator also provide protection against dusts/mists if appropriate filter/cartridge/canister is selected.

1.4 Prepare chemical(s) for application on seeds as per the SOP

1.4.1 Introduction

Pesticide is a group which contains:

- insecticides,
- fungicides,
- herbicides,
- rodenticides,
- garden chemicals,
- wood preservatives
- household disinfectants

Pesticides have been an essential part of agriculture as they protect crops and livestock from pest infestations and gives reduction for many decades. It is poisons and can be hazardous.

1.4.2 Selecting Pesticides

- The initial step is to accurately select a pesticide. If the pest is misidentified, fertilizer and pesticides applicator will not be able to choose effective pesticide or other management strategy.
- If a pesticide is needed, select one that is effective against pest and also poses the least risks to human health and environment.
- A good source of information for identifying effective, least-toxic methods and pesticides for use against specific pests is to consult with authorities.
- When shopping for pesticide, fertilizer and pesticides applicator should consult the label to be sure the target pest and site is listed.
- Before purchasing pesticide, fertilizer and pesticides applicator should check the label to be sure it is appropriate to use on plants or treatment site.
- When choosing pesticides, fertilizer and pesticides applicator should remember that most pesticides only control certain stages of pest while many insecticides kill only larval stage and not the eggs or pupae.

1.4.3 Instructions to be followed before preparing solution

Fertilizer and pesticide applicator should clearly follow instructions before preparing the pesticides solution which covers:

- Pesticide labels often do not contain directions for mixing small quantities of spray, so accurate measurements of ingredients are important when mixing pesticides.
- The concentration of active ingredient in the spray is critical to its effectiveness, so use electronic scale to measure dry ingredients.

- Label directions for mixing and applying pesticides can be rate per volume of water or rate per area of land.
- Pesticides that are mixed by concentration generally have directions for application.
- Some insecticide application directions may state to apply until spray runs off while some herbicide application directions may state to apply only enough spray material to wet leaves uniformly.
- If pesticide mixing directions state an amount of formulated material per volume of spray, then adjust the amount of pesticide to the volume of water which needs to be mixed.

- Mix only enough spray to cover the area, as for spraying 2000 square feet with 1 gallon for 1075 square feet, then it should be 2000/1075 which gives number of gallons of spray.
- Densities of solid pesticides varies with formulation and amount of shaking or settling in the package during shipping and in storage, so electronic scale will make sure of correct weight of the dry product usage.

1.4.4 Estimation and Procedure of Pesticide Solution

Fertilizer and pesticide applicator should clearly know how much solution is prepared and applied in particular area. For this, they have to clearly follow and calculated the solution usage with necessary preparation procedure which should cover:

a) Requirement of Spraying Solution for 1 acre of area

- Required insecticide spray solution: 180 litres
- Required concentration (%): 0.076 (leaf roller and wingless grass hopper), 0.15 (hairy caterpillar) and 0.2 (mealy bug)

- Required quantity of commercial insecticide formulation: 180 ml (leaf roller and wingless grass hopper), 360 ml (hairy caterpillar) and 473 ml.

b) Preparation of Recommended Concentration of Spray Solution

- Add the required quantity of commercial insecticide to small quantity of water and adjust the final volume of the spray solution to 180 litres by adding water.

- Stir the spray solution for uniform distribution of the insecticide in the diluent (water).
- Spray the insecticide solution using a sprayer till the leaves become wet.
-

Note

- Many pesticides applied as spot treatments are mixed with water at specific percent solution, so to create such solution, seed processing worker needs to convert the percentage of concentrate in final solution to decimal before calculating the amount of concentrate to mix with water.
- To convert percentages to decimals, divide the percentage number by 100 to get the decimal which is same as moving the decimal point two places to left and adding zeroes as needed.
- Pesticides and other ingredients of liquid mixture should be considered a part of total solution.
- If the seed worker want to make 100 gallons of 10% solution of pesticide solution in water, in such case, he/she needs to add 10 gallons of pesticide concentrate to 90 gallons of water.
- If seed worker adds 10 gallons of pesticide to 100 gallons of water, then solution would be too dilute.

1.4.5 Prepare stock solutions and solution of pesticides/ fertilizers as instructed

Stock solution

- A stock solution is a concentrated solution that will be diluted to some lower concentration for actual use.
- These are used to save preparation time, conserve materials, reduce storage space, and improve the accuracy with which working lower concentration solutions are prepared.

Preparing Stock Solutions

- A stock solution is prepared by weighing appropriate portion of pure solid or measuring appropriate volume of pure liquid and then diluting to known volume.
- It can be done based on the required concentration unit.
- The stock solutions are mixed as per fertilizer injector ratio where every injector delivers certain amount of stock solution for each increment of irrigation water which passes through injector. The injector ratio shows the concentration of stock solution that is required to deliver a particular rate of fertilization.
- Injectors like Hozon, Smith Measuremix have fixed injector ratio whereas injectors like Anderson, Dosmatic, Dosatron, M-P Mixer Proportioner have adjustable ratios.
- Many growers prefer injectors with adjustable ratios so that different fertilizer rates can be applied to crops with different nutrient requirements.
- Preparing of fertilizer stock solutions is of great importance as it involves accurately determination of amount of fertilizer that is needed to mix the stock solutions of fertilizers.
- Many manufacturers of commercial fertilizers and fertilizer injectors show information on fertilizer bags.
- Without recourse to tables/bags, pesticide and fertilizer applicator can use formulas to calculate the amount of fertilizer required.
- In this, if pesticide and fertilizer applicator knows the rate of fertilization (in ppm N), percentage of nitrogen in fertilizer, and injector ratio, then he/ she can calculate by using the formula:
- Amount of fertilizer to make 1 volume of stock solution = Desired concentration in parts per million x Dilution factor
- The formula allows to calculate the amount of fertilizer required to mix the stock solutions.

- ❖ With this, pesticide and fertilizer applicator can use it with any fertilizer injector and with all common units of measurement.

Preparation of liquid fertiliser stock solutions

- ❖ Liquid fertiliser is commonly used on its own or as a supplement to base fertiliser in horticulture.
- ❖ As a range of proprietary liquid feeds, it is cheaper to mix own recipe from single nutrients.
- ❖ The formula can be tailored to specific crop and the ratio of nutrients adjusted to growth stage and time of year.
- ❖ It is convenient to make concentrated batches of fertiliser and dilute them through proportioning device and such batches of concentrated nutrient solution are called as stock solutions.

Steps to prepare stock solution

There are certain steps that describe the preparation of stock solution:

Mixing

- ❖ Fertilisers that are different in amounts can be dissolved in water.
- ❖ Many fertilisers dissolve more readily in warm or hot water.
- ❖ When making concentrated stock solutions it is required to know the solubility of chemicals involved.
- ❖ Pesticide and fertilizer applicator should know that, if stock solution is heated to dissolve the components, in such case, fertiliser may crystallise on cooling of solution, soo to avoid it, dilute the solution or use it immediately.

Calculating liquid feed solutions

- ❖ For this, pesticide and fertilizer applicator should remember 1ppm = 1mg/L (1 part per million = 1 milligram per litre) and 1 milligram = 1/1000 of a gram (1000 milligrams = 1 gram).
- ❖ Pesticide and fertilizer applicator needs to decide which fertilisers are to be used for nitrogen, phosphorus and potassium sources.

1.5 Prepare tools and equipments for manual application of chemical(s) on seeds

1.5.1 Equipments

The idea of proper pesticide application cannot be obtained, if good quality protection equipments are not used. Fungicides are applied to seed prior to planting to provide effective protection against many seed and soil-borne plant pathogens. Chemical treatment guards against the various seed rots and seedling blights that occur during storage or after planting. It is not usually a "cure-all" and will not provide disease protection throughout the growing season after the plants become self-sufficient.

Insecticides are often applied to seed to control or reduce insect damage to seed during storage and, to a lesser degree, to prevent damage from such insects as wireworms and seed corn maggots in the soil.

A well designed machine will be efficient as far as pesticide distribution and delivery to target in minimum time with minimum wastages is concerned, so machines should be tested to ascertain:

- Efficiency
- Reliability
- Life
- Mode of operation
- Minimum field problems

The equipment so used should meet certain minimum requirements of performance, efficiency and reliability, so such equipments should have standard specifications which includes:

- Material of construction
- Dimensions
- Ergonomics
- Stability
- Safety
- Interchangeability
- Performance
- Strength, Reliability
- Workmanship, Finishing

1.5.2 Types of Equipments

Sprayers

SPRAYERS (Hydraulic energy)

Manually operated	Powered operated
1. Syringes, slide pump	1. High pressure sprayer (hand carried type)
2. Stirrup pumps	2. High pressure trolley/ Barrow mounted
3. Knap sack or shoulder-slung:	3. Tractor mounted/ trailed sprayer

- Lever operated K.S. sprayer
- Piston pump type
- Diaphragm pump type

4. Compression sprayer
   - Hand compression sprayer
   - Conventional type
   - Pressure retaining type

4. High pressure knap sack sprayer

5. Stationary type
   - Foot operated sprayer
   - Rocker sprayer

5. Air craft, aerial spraying

SPRAYERS (Gaseous energy)

Manually operated	Powered operated
1. Hand held type	1. Knap sack, motorized type
	2. Hand/ Stretcher carried type
	3. Tractor mounted

SPRAYERS (Centrifugal energy)

1. Hand held battery operated ULV sprayer.
2. Knapsack motorized type
3. Tractor/ vehicle mounted ULV sprayer
4. Aircraft ULV sprayer

Dusting Equipments

Manually operated	Powered operated
1. Plunger duster	1. Knapsack motorized duster
2. Bellow duster	2. High pressure trolley/ Barrow mounted
3. Rotary duster: • Belly mounted model • Shoulder-slung model	3. Tractor mounted/trailed duster
	4. Aircraft

Granule Applicator

Manually operated	Powered operated
1. Broad-casting tins	1. Knapsack motorized type
2. Knapsack Rotary granule	2. Tractor mounted/ trailed duster
	3. Aircraft

Spraying nozzle

Spraying nozzle is a device for emitting spray liquid, breaking it up into small droplets and throwing droplets away from the nozzle orifice. Different designs of nozzle are used to produce required droplet size spectrum. To break the liquid into droplets, energy is required. The spray nozzles are classified as:

- Hydraulic energy nozzles
- Gaseous energy nozzles
- Centrifugal energy nozzles
- Thermal energy nozzles

Seed Treatment Equipment

There are many commercial seed treaters that are designed to give accurate measured quantities of pesticides to a given weight of seed. Normally, there are three commercial types of seed treaters such as:

- dust treaters
- slurry treaters
- direct treaters

Dust Treater

- In this, amount of seed which flows in weigh pan is controlled by opening or closing the gates of feed hopper using hand wheel on side of hopper.

- The scale on hopper shows how far the gates are open.
- In this, gates should be open to whatever number of inches it takes to keep the weigh pan filled to the required number of pounds per dump as it tilts in either direction.
- The number of pounds per dump is adjusted by correctly setting the counterweight up or down on the counterweight arm.
- The correct amount of powder is applied to the seed flow, a preliminary test must be made in which a given number of pounds of seed is run by feeder.
- During this run, measuring cup provided with feeder should be used to catch the powder as it comes off the vibrator.
- After given amount of seed has run through, powder should be weighed to determine how much is being applied to that amount of seed.
- In this, vibrator speed can be adjusted accordingly and more tests should be run until proper setting of the vibrator speed is determined for correct coverage.

Slurry Seed Treater

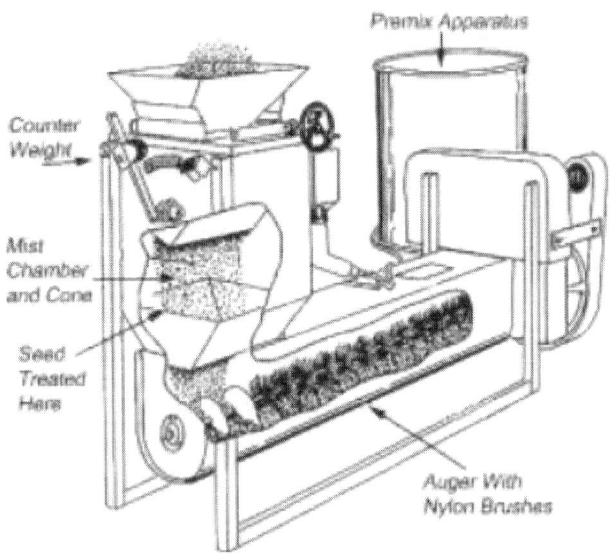

- The slurry treatment principle uses suspension of wettable powder treatment material in water.
- The treatment material applied as slurry is accurately metered through a simple mechanism composed of slurry cup and seed dump pan.
- The cup in this, introduces given amount of slurry with each dump of seed into a mixing chamber where they are blended.

- In this, metering principle is same as in direct, ready-mix or fully automatic treaters as introduction of fixed amount of slurry to given weight of seed.
- To obtain a given dump weight, slurry treaters are equipped with seed gate that controls seed flow to dump pan.
- With proper seed gate setting, constant dump weight for given can be obtained.
- The amount of treatment material applied is adjusted by slurry concentration and size of slurry cup or bucket.
- As dump pan fills, point is reached where it over-balances the counter weight and dumps into the mixing chamber which brings alternate weighing pan in position to receive the inflow of seed and activates a mechanism to add a cup of slurry to the mixing chamber.
- The mixing chamber in this is fitted with an auger type agitator that mixes and moves seed to the bagging end of the chamber.
- In tthis, speed of auger is important as at slow speeds, more uniform distribution is obtained.
- Slurry tanks have 15 to 35 gallon capacities, depending upon the size of the treater.
- The tanks are equipped with agitators that mix the slurry in the tank and keep it suspended during operation.
- It is important that the powder be thoroughly suspended in water before treating.
- If treater has been idle for any period of time, sediment in bottom of slurry cups needs to be cleaned out.

Direct Treater

- It is of two types; Panogen and Mist-O-Matic treaters.
- In panogen treater, there is a small treatment cup, which operates from rocker arm directly off the seed dump pan and out to small reservoir.
- The rocker arm meters one cup of treatment with each dump of seed pan.
- In panogen seed treater, fungicide flows through a tube to the head of revolving drum seed mixing chamber which flows with seed from dumping pan which gets distributed over seed by rubbing action of seed that passes through revolving drum.
- In panogen seed treater, desired treating rate is obtained by size of treatment cup and adjustment of seed dump weight.
- Treatment cup sizes in panogen seed treater are designated by treating rate in ounces and not by actual size as 3/4 ounce cup applies 3/4 ounce (22.5 cc) of treatment per bushel with six dumps per bushel.

- The actual size of treatment cup is approximately 3.75cc.
- The mist-o-matic treater applies treatment as mist directly to the seed.

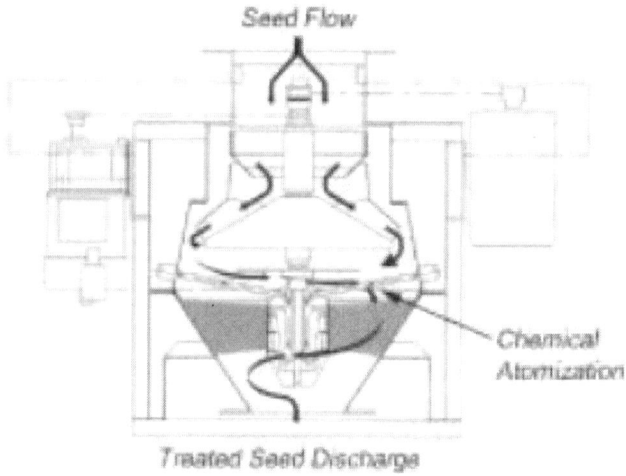

- In mist-o-matic treater, the metering operation of treatment cups and seed dump is similar to panogen treater.
- In mist-o-matic treater, the cup sizes are designated by number of cc's they actually deliver as 2 ½ , 5, 10, 20 and 40.
- The treater in mist-o-matic treater is equipped with large treatment tank, a pump and a return that maintains the level in the small reservoir from which the treatment cups are fed.
- After metering, treatment material in mist-o-matic treater flows to a rapidly revolving, fluted disc mounted under a seed-spreading cone.
- The disc breaks droplets of treatment solution in fine mist and sprays it outward to coat seed falling over the cone through the treating chamber.
- Just below the seed dump, in mist-o-matic treater, there are two adjustable retarders designed to give continuous flow of seed over cone between seed dumps.
- The desired treating rate in mist-o-matic treater is obtained by selection of treatment cup size and proper adjustment of the seed dump weight.

1.5.3 Use of equipments as per instruction manual

Before using and handling pesticide/fertilizer equipment, seed applicator should clearly follow the instructions shown below:

- Read instructions manual of the pesticide and its equipment.
- Check the spraying equipment and accessories which are to be used.
- Ascertain that all components are clean, especially filling and suction strainer, sprayer tank, cut off device and nozzle.
- Replace worn out parts like 'O' ring, seal, gasket, worn out nozzle tip, hose clamps and valves.
- Test the sprayer and ascertain whether it pumps the required output at rated pressure.
- Check the nozzle spray pattern and discharge rate.

Preparing the sprayer

- Before spraying operation is started, rinse out the sprayer; remove and clean all nozzles, nozzle screens and strainers.
- Make sure strainers and nozzle screens are 50 mesh or larger when wettable powders are used.
- All nozzles should be of same type, size and fan angle for most applications.
- Check all lines, valves, seals and the tank after filling the sprayer with water and during calibration to be sure there are no leaks in the spray system.
- For safety, replace weather-cracked or worn hoses.
- Adjust the nozzle height and spacing as suggested by the nozzle manufacturer or as specified on the pesticide label.
- Only water that is clean enough to drink should be used in the sprayer.
- A small amount of silt or sand in the water rapidly wears pumps and other parts of the sprayer system.
- Operating a pump dry or with restricted inlet may damage the pump, so do not operate pumps at speeds or pressures higher than the manufacturer recommends.

- ❖ Power take-off pumps should be restrained from rotating by chains or torque bars.
- ❖ Do not use ground speeds that are too high for existing field conditions.
- ❖ Booms bouncing up and down or back and forth can cause application rates to vary by 50 percent, so such bouncing can damage the spray booms or frame of sprayer.

1.6 Switch on 'Seed Treating Barrel' or 'Mixer' in mechanized seed processing units as per SOP

- ❖ Seed treatment is a means of preventing or reducing the risks from number of soil borne and seed borne pathogens or insects.
- ❖ For large amount of seeds, seed treatment involves mechanized seed processing units.
- ❖ The mechanized seed processing unit uses seed treatment mixer drum to treat the seeds.
- ❖ Seed processing mixer drum consists of frame, handle and cylindrical drum.
- ❖ In this, the drum is cylindrical which is mounted on tri-pod angle iron frame with three piece of mild steel flat which are welded in drum for uniform mixing.
- ❖ To use the mixer, seed processing worker needs to add chemicals in the drum along with little water and should close the lid of drum tightly to allow the drum to rotate for 20 to 25 revolutions.
- ❖ For rotating the drum, seed processing worker needs to switch ON the mixer, so that the mixer drum should rotate for 1-2 minutes.
- ❖ After rotation, the processing worker will open the drum lid and take the treated seed out in separate bag/container.

2

Apply Chemicals Manually

2.1 Fill prepared chemical(s) manually in appropriate tool or equipment safely

2.1.1 Filling prepared chemical(s) manually in appropriate tool or equipment safely

- Once the required chemical is prepared, then pesticide applicator/ processing worker can store the pesticide in a can or can directly fill in the pesticide equipment to use for seeds.
- The prepared chemical/pesticide should be handled by using goggles and gloves to safeguard from pesticide exposure.

Filling of sprayer

- Before the prepared chemical is mixed in sprayer tank, seed processing worker or pesticide applicator should check the sprayer and calibrate it with water.
- He/she should use wooden paddle or stick to mix water-dispersible powders with small quantity of water to form a smooth paste to be added to sprayer tank.

- ❖ The processing worker should use more water to rinse the spray container.
- ❖ After cleaning the prayer, pesticide applicator should pour the suspension in spray tank using a strainer or filter funnel to prevent dirt from entering as when suspension is not filtered, in such case, nozzle tip will get blocked during spraying.
- ❖ Spray tank should not be more than three-quarters filled and the remaining space should be left for compressed air.

- ❖ Spray tank usually has mark that shows the required chemical level.

2.1.2 Precautions

While Preparing spray solution, processing worker or pesticide applicator should:

- ❖ Always use clean water.
- ❖ Use protective clothings viz., hand gloves, face masks, cap, apron, full trouser, etc. to cover whole body.
- ❖ Always protect your nose, eyes, ears, hands, etc. from spill of spray solution
- ❖ Read instructions on pesticide container label carefully before use.
- ❖ Prepare the solution as per requirement.

- Granular pesticides should be used as such.
- Avoid spilling of pesticides solutions while filling the spray tank.
- Always use recommended dosage of pesticide.
- No activities should be carried out which may affect your health.

2.2 Apply chemical(s) manually on seeds using appropriate tool or equipment as per SOP

Pesticide application plays an important role in pest management. Proper technique of application of pesticide and the equipment used for applying pesticide are vital to the success of pest control operations. The application of pesticide is not merely the operation of sprayer or duster. It has to be coupled with a thorough knowledge of the pest problem. The use of pesticides involves knowledge not only of application equipment, but of pest management as well.

The purpose of pesticide application technique is to cover the target with maximum efficiency and minimum efforts to keep the pest under control as well as minimum contamination of non-targets.

The application techniques should be target oriented so that safety to non-targets and environment is ensured. For this, proper selection of application equipment, knowledge of pest behaviour and skillful dispersal methods are important. The complete knowledge of pest problem is important to define the target which includes location of pest.

Different pesticides are used for controlling different pests, as insecticides are applied for insect pests, fungicides are applied for crop diseases, herbicides are for weeds etc, for protecting crop losses. The application of pesticide is successful when applied at susceptible stage of pest. If timing of pesticide application is carefully considered and followed, then the results will be good pest control and economy, so large area treatment, careful selection of equipment becomes necessary so that in available time, the area could be treated.

For small amount of seeds to be treated, seed processing worker can do manual seed treatment by using certain brush, so that chemical can be equally brushed on every seeds.

2.3 Take sample of treated seeds, label them and send to lab for quality check

2.3.1 Taking treated seed sample

Seed Quality

Seed quality is the possession of seed with required genetic and physical purity which carries physiological soundness and health status. The major seed quality characters are:

Physical Quality

- It is the cleanliness of seed from other seeds, debris, inert matter, diseased seed and insect damaged seed.
- The seed with physical quality should be of uniform size, weight, and colour.
- The seed should be free from stones, debris, and dust, leafs, twigs, stems, flowers, fruit well without other crop seeds and inert material.
- Seed should be devoid of shriveled, diseased mottled, moulded, discoloured, damaged and empty seeds.
- The seed should be easily identifiable as species of specific category of specific species.
- Lack of quality character in seed will indirectly influence the field establishment and planting value of seed.
- The quality character of seed could be obtained with seed lots by proper cleaning and grading of seed after collection and before sowing / storage.

Genetic purity

- It is true to type nature of seed.
- The seedling / plant / tree from the seed should resemble its mother in all aspects.
- The quality character of seed is important for achieving desired goal of raising the crop either yield or for resistance or for desired quality factors.

Physiological Quality

- It is the actual expression of seed in further generation / multiplication.
- Physiological quality characters of seed have seed germination and seed vigour.
- The liveliness of seed is known as viability.
- The extent of liveliness for production of good seedling or ability of seed for seedling with normal root and shoot under favorable condition is known as germinability.
- Seed vigour is energy or stamina of seed in producing elite seedling which is the sum of all seed attributes that enables its regeneration under any given conditions.
- Seed vigour shows level of performance of seed or seed lot at germination and seedling emergence.
- Seed that do well at sowing are termed as quality seed and based on degree of performance in production of elite seedling, it is categorised as high, medium and low vigour seed.

- The difference in seed vigour is differential manifestation of deteriorative process which oocurs in seed before ultimate loss of ability to germinate.
- Difference in seed vigour is expressed in rate of emergence, uniformity of emergence and loss of seed germination.
- The viable seeds needs no germination but all germinable seed will be viable.
- The vigourous seeds will be germinable but all germinable seed need not be vigourous. Physiological quality of seed is achieved by proper selection of seeds used for sowing and by caring for quality characters during extraction, drying and storage.
- Seed with good vigour is preferable for raising a good plantation as the fruits, the economic come out are to be realized after several years.
- The selection of seed based on seed vigour is important for raising perfect finalize plantation.

Seed Health

- The health status of seed is the absence of insect infestation and fungal infection, in or on the seed.
- Seed should not be infected with fungi or infested with insect pests as these will reduce the physiological quality of seed and also the physical quality of the seed in long term storage.
- The health status of seed includes deterioration status of seed expressed by low vigour status of seed.
- The health status of seed influences the seed quality characters directly and warrants their soundness in seed for production of elite seedlings at nursery / field.

Seed sample

- Once the seed is treated with chemicals, seed processing worker should collect the few treated seeds sample in order to check the seed quality.
- Treated seed sample can be checked for quality at various government and private laboratories by paying certain nominal amount.
- The seed sample should be taken in clean dry container which shall be closed sufficiently tight to prevent leakage and entrance of moisture.

- ❖ Seed growers knows that treated seeds contain certain herbicide or insecticide, so they needs to carefully handle treated seeds by using appropriate personal protective equipment as mentioned in SOP, since wearing of appropriate PPE will help in reducing or preventing seed workers from pesticide exposure while working with treated seed.
- ❖ Seed processing worker while wearing required gloves, should use their hands to collect sample of treated seeds from the container.

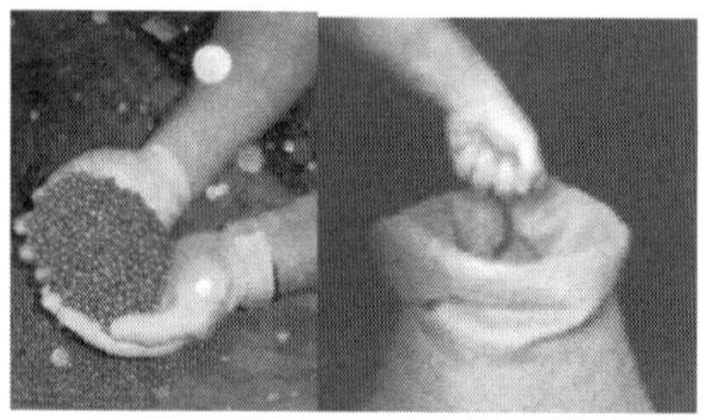

Types of Seed Samples

Seed testing samples used in Seed Testing Laboratory includes:

- ❖ Service sample: It is a sample which is received from other than seed certification agencies and seed inspectors
- ❖ Certified sample: It is a sample which is received from certification agencies or officers
- ❖ Official sample: It is a sample which is received from seed inspectors.

Handling of treated seed

- ❖ Comply with all label instruction for seeding operations.
- ❖ Use high quality dust-free seed.
- ❖ Wear gloves when handling treated seed and when planting.
- ❖ Wear dust masks when opening and/or emptying treated seed packaging to minimise contact with dust.
- ❖ Avoid soaking pre-treated seed as it could become contaminated with chemicals.
- ❖ Properly dispose of any spillage to prevent exposure to humans, animals, or the environment.
- ❖ Ensure emergency contact information is easily accessible.

2.3.2 Labelling of treated seed sample

- ❖ Once the seeds are kept in seed sample container and are sealed with required material, they needs to be properly labeled.
- ❖ Seed sample containers for analysis shall be properly labeled and addressed with following information:

(a) serial number;

(b) name of sender with official designation, if any;

(c) name of person from whom the sample has been taken;

(d) date and place of taking the sample;

(e) kind or variety of the seed for analysis;

(f) nature and quantity of preservative, if any, added to the sample;

Sending Seed Sample to Laboratory

- Once the treated seed samples are marked with proper labeling, they need to be sent for analysis with proper packing, fastening and sealing where:
  - Stopper shall initially be securely fastened so to prevent leakage of containers in transit.
  - Sample container shall be completely wrapped in fairly strong thick paper where the ends of paper shall be neatly folded in and affixed using gum or adhesive.
  - The paper cover shall be secured using strong twine or threads, both above and across sample seed container where twine or thread needs to be fastened on paper cover by sealing wax with distinct and clear impressions of seal of sender, at top, bottom and sides of packet.

2.3.3 Quality check of seeds

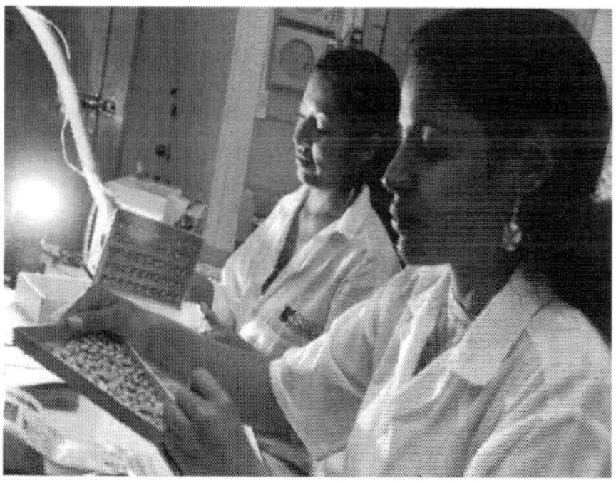

- Seed Testing Laboratory will analyze seed samples as per prescribed procedure and deliver the Seed Analysis Report to user/Certification Agency within 30 days from the date of receipt of samples.
- Once the seed samples are received in testing lab, the respective samples get registered and are assigned with laboratory test numbers and code for subsequent analysis.
  - The procedure for seed quality testing involves:

- Conducting of moisture test, soon after registration of seed samples.
- Need of working samples from division of submitted sample for various tests using mechanical divider, random cups, spoon or by modified halving method.
- Calculating weight of working samples for purity analyses with 2500 seeds or maximum of 1000 gm of sample weight.
- Separating of working sample in selected component and rest component, where fractions are weighed and weight of former sample is calculated as percentage of both.

2.4 Send treated seeds for further processing

- Once the seeds have been tested, now an important tasks is to interpret the test results obtained which involves consideration of purity, weed seed content, viability, hard seed content, disease organisms carried by seed and its resistance or susceptibility to organisms in soil.
- The seed testing report will show the further action to be taken by seed processing worker for further processing.
- As per the seed standards details for genetic purity, the tested seed lots should conform the following minimum standards for genetic purity which shows;
    - Foundation/Certified seeds 99.00
    - (i) Varieties, composites, synthetics, Multilines 98.00
    - (ii) Hybrids 95.00
- Apart from this, the laboratories will conduct grow-out test in order to determine the genetic purity of seed lot before further processing of testing seeds.
- On receipt of Seed Analysis Report and results of grow-out test, if seed lot met prescribed standards, then the seeds can be treated as certified and under such case, such seed lot can be send for further packing, sealing and for certification.
- Containers used for packing of tested/certified seeds shall be durable and free from defects.
- The testing lab/agency have the authority to refuse testing seeds for certification, if the testing seeds lot does not conform to minimum standards as prescribed for particulars crop.

3

Apply Chemicals Mechanically

3.1 Fill 'Seed Treating Barrel' or 'Mixer' with prepared chemical(s) safely as per SOP

- Chemical treatment guards against various seed rots and seedling blights that occur during storage or after planting.
- Once the seed processing worker prepares the required solution to treat seeds in barrel or mixer, he/she will fill the required solution as per instructions.
- Drum mixers, are gallon drums that rotate in order to mix chemicals and materials in tanks.
- Chemicals for treating seeds can be directly be used or can be prepared as per type of seeds to be treated.
- Seed processing worker should fill the drum/mixer with one-half to two-third full of seeds.
- The required quantity of chemicals to be mixed with seeds should be carefully poured inside the drum.
- The mixer drum consists of frame, handle and cylindrical drum which is mounted on tri-pod angle iron frame with three piece of mild steel flat which are welded in drum for uniform mixing.
- To work with mixer containing seeds, seed processing worker needs to add slight water in chemicals for smooth mixing.

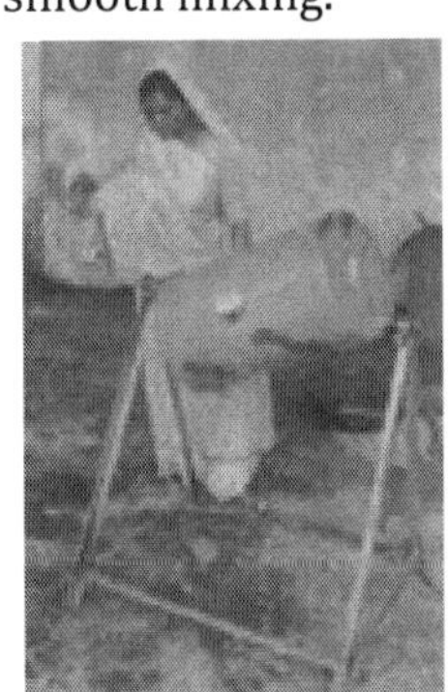

- After adding chemical and water in mixer, the lid of drum is made tight and the drum is allowed to rotate for 20 to 25 revolutions for proper mixing.

3.2 Operate machine after feeding seeds in it at appropriate rate as per SOP

3.2.1 Seed Rate

- As seeds are important for farming and for required field crop, so proper estimation of seed is required for particular field and crop.
- Technology has modernized much of farming's day-to-day operations, but without steady supply of high-quality seed, yields and crop quality would be greatly decreased.
- Seed quality plays an important role in production of agronomic and horticultural crops.
- Seed characteristics like trueness to variety, germination percentage, purity, vigor and appearance are important to seed workers in planting crops.
- Numbers of plants in crop depends on seed rate, seed viability, percentage seedling emergence and plant survival, however, plant numbers can often vary widely without appreciably affecting yield in irrigated crops as plants produce tillers, which themselves produce leaves, spikes and grain.
- As seeds are of different sizes and shapes from micro-miniature orchid seed, to small particle of dust and to hat of gargantuan double coconut seed, which is 1 feet long and weighs many pounds, so seed worker needs to properly assess the amount of seeds required.
- The shapes of the seed varies from simple, round tobacco seed to complex, aerodynamic, winged maple seed, so proper estimation is also required in such.
- Some seeds are hairy like cotton or thistle while others are like grass, have long spikes while some have thorns.

Factors to consider for Seed Rate

The required number of crops/plants per unit area can be decided by calculating the seed rate which depends on:

- Plant spacing
- Plant population
- Test weight
- Germination percentage
- Weight of seed
- Plant type; tall or short

Calculating Seed Rate

To calculate seed rate, seed processing worker should apply the following formula:

$$\text{Seed rate (kg/ha)} = \frac{\text{Plant population (per ha)} \times \text{No. of seeds/hill} \times \text{Test weight (g)} \times 100}{1000 \times 1000 \times \text{Germination percentage (\%)}}$$

3.2.2 Operating Seed Treating Barrel/Mixer

- ❖ Once the seed rate is calculated, required quantity of seeds are put in seed treating mixer drum.

- ❖ Seed processing mixer drum consists of frame, handle and cylindrical drum which is mounted on tri-pod angle iron frame with three piece of mild steel flat which are welded in drum for uniform mixing.
- ❖ The drum has a lid from which seeds and chemicals are poured inside the drum.
- ❖ Once the seed processing worker finish off with addition of water, seeds and chemicals in required quantities, he/she will close the drum lid.
- ❖ As mostly, the seed treating mixer is manually operated using a rod, so seed processing worker after sealing the lid will rotate the drum manually for 2-3 minutes for proper mixing.
- ❖ If the seed treating mixer is automatic, then seed processing worker will switch ON the mixer supply, so that the mixer drum should rotate for 2-3 minutes.
- ❖ After rotation, the processing worker will open the drum lid and take the treated seed out in separate bag/container.

3.3 Take samples of treated seeds, label them and send to lab for quality check

3.3.1 Preparing Seed sample

- Once the required amount of seeds are treated with chemicals, then seed processing worker should collect few treated seeds and should make samples of treated seeds to be send for quality check in labs.
- Treated seed sample can be checked for quality at various government and private laboratories by paying certain nominal amount.
- The seed sample should be taken in clean dry container which shall be closed sufficiently tight to prevent leakage and entrance of moisture.

- Seed growers knows that treated seeds contain certain herbicide or insecticide, so they needs to carefully handle treated seeds by using appropriate personal protective equipment as mentioned in SOP, since wearing of appropriate PPE will help in reducing or preventing seed workers from pesticide exposure while working with treated seed.
- Seed processing worker while wearing required gloves, should use their hands to collect sample of treated seeds from the container.

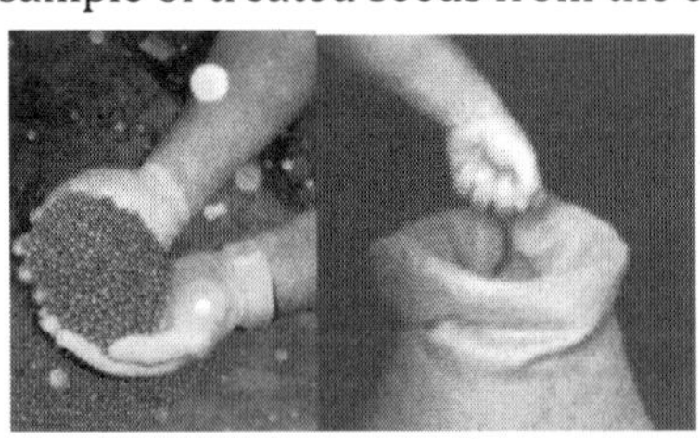

- Seed processing worker can also use sleeve type sampling stick for taking seed sample which comprise of inner and outer tube having partition walls.

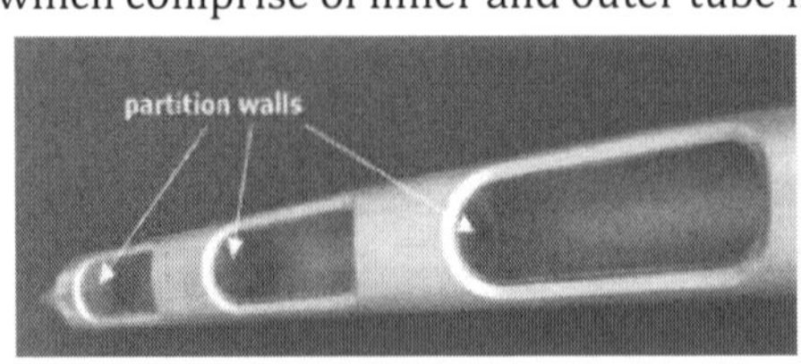

Types of Seed Samples

- As in the seed testing laboratory, only small portion of seed lot can be tested, so it is important for seed processing worker to keep the small seed lot for testing.
- Seed testing samples used in Seed Testing Laboratory will include, service sample which is received from other than seed certification agencies and seed inspectors, certified sample which is received from certification agencies or officers or official sample which is received from seed inspectors.

3.3.2 Labelling of seed sample

- Once the seeds are kept in seed sample container and are sealed with required material, they need to be properly labeled.
- Seed sample containers for analysis shall be properly labeled and addressed with information related to serial number; name of sender with official designation; name of person from whom the sample has been taken; date and place of taking sample; kind or variety of seed for analysis; nature and quantity of preservative added to seed sample.
- Once the seed samples are marked with proper labeling, they need to be sent for analysis with proper packing, fastening and sealing.

3.3.3 Quality check of seeds

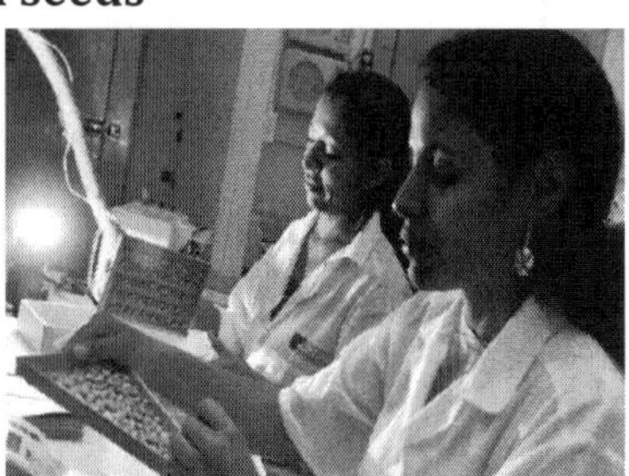

- Seed Testing Laboratory will analyze seed samples as per prescribed procedure and deliver the Seed Analysis Report to user/Certification Agency within 30 days from the date of receipt of samples.

- ❖ Once the seed samples are received in testing lab, the respective samples get registered and are assigned with laboratory test numbers and code for subsequent analysis.
- ❖ Once the seed samples are entered, they are mixed and divided on mechanical divider before sending it for purity analysis.

- ❖ For this, the sample needs to run through mechanical divider 3-4 times to mix properly.
- ❖ The submitted sample will be divided till working sub-sample of required size is obtained as per ISTA rule.

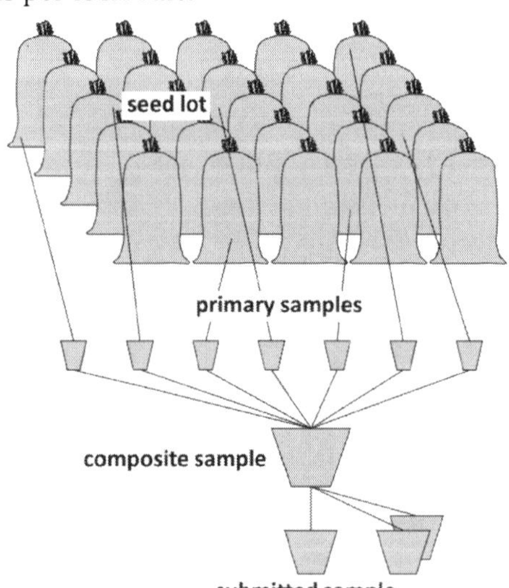

- Also, the seeds in single pan need to be set aside after every division while other half should run through divider again to get correct working sample size.
- In such case, if weight of sub sample is less than required sample size, then it can be increased by adding the part from sample, which was set aside.
- If sample is too large, such can be reduced by repeated mechanical division.
- When the required seed sample is ready, then seed processing worker will involve in seed quality testing procedure which involves:
  - Conducting of moisture test, soon after registration of seed samples.
  - Need of working samples from division of submitted sample for various tests using mechanical divider, random cups, spoon or by modified halving method.
  - Calculating weight of working samples for purity analyses with 2500 seeds or maximum of 1000 gm of sample weight.
  - Separating of working sample in selected component and rest component, where fractions are weighed and weight of former sample is calculated as percentage of both.

Seed lot sampling in containers for 15-100 kg of seeds

Containers in seed lot	Minimum number of primary samples
1 - 4 containers	3 primary samples from each container
5 - 8 containers	2 primary samples from each container
9 - 15 containers	1 primary samples from each container
16 - 30 containers	15 primary samples from the seed lot
31 - 59 containers	20 primary samples from the seed lot
60 or more containers	30 primary samples from the seed lot

3.4 Send treated seeds for further processing

- Once the seeds have been tested, now an important tasks is to interpret the test results obtained which involves consideration of purity, weed seed content, viability, hard seed content, disease organisms carried by seed and its resistance or susceptibility to organisms in soil.
- The seed testing report will show the further action to be taken by seed processing worker for further processing.
- As per the seed standards details for genetic purity, the tested seed lots should conform the following minimum standards for genetic purity which shows;

- Foundation/Certified seeds 99.00
- (i) Varieties, composites, synthetics, Multilines 98.00
- (ii) Hybrids 95.00

- Apart from this, the laboratories will conduct grow-out test in order to determine the genetic purity of seed lot before further processing of testing seeds.
- On receipt of Seed Analysis Report and results of grow-out test, if seed lot met prescribed standards, then the seeds can be treated as certified and under such case, such seed lot can be send for further packing, sealing and for certification.
- Containers used for packing of tested/certified seeds shall be durable and free from defects.
- The testing lab/agency have the authority to refuse testing seeds for certification, if the testing seeds lot does not conform to minimum standards as prescribed for particulars crop.

4

Achieve Productivity and Quality Standards

4.1 Apply accurately measured quantities of chemical(s) as described in SOP

Using right amount of pesticide and fertilizer will minimize the potential for personal, property, or environmental damage and maximizes consumers return.

4.1.1 Preparing required pesticide solution

Seed processing worker should clearly know how much solution needs to be prepared and applied in particular area.

For this, he/she has to clearly follow and calculated the solution usage with necessary preparation procedure which should cover:

The Spraying Solution required for 1 acre of area should have:

- insecticide spray solution of around 180 litres
- percentage concentration of 0.076 for leaf roller and wingless grass hopper, 0.15 for hairy caterpillar and 0.2 for mealy bug
- quantity of commercial insecticide formulation as 180 ml for leaf roller and wingless grass hopper, 360 ml for hairy caterpillar.

Preparing Recommended Concentration of Spray Solution

- Add required quantity of commercial insecticide to small quantity of water and adjust the final volume of spray solution to 180 litres by adding water.

- Stir spray solution for uniform distribution of insecticide in diluent (water).
- Spray the insecticide solution using a sprayer till the leaves become wet.

4.1.2 Applying required amount of chemicals

- Once the measured quantity of chemical solution is prepared, seed processing worker will not apply such solution on respective seeds.
- Seed processing worker along with required quantity of seeds can add the required amount of chemical in seed mixer for proper mixing.
- With required chemicals and seed quantity, seed processing worker uses sprayer for proper chemical application on seeds.

4.2 Ensure that required amount of chemical is applied uniformly over given amount of seed

- Normally chemicals on seeds can be applied using certain tool and instrument.
- If the seed processing worker has filled the required amount of chemical in the tool or equipment for spreading on seeds and if the required chemicals is not placed or sprayed evenly, then in such case, the seed worker needs to check their tool/equipment.
- Calibration is the process of measuring and adjusting the amount of pesticide which an equipment will apply over targeted area.
- Calibration is an initial step which makes certain that the tool/equipment used is applying pesticide uniformly and at correct rate or not.
- Calibrating tool will save money by not wasting pesticides from over application as over-applying pesticides can result in excess residues on or in plants, soil, and surface or groundwater.

Unit 3: Pack and Label Seeds

1

Undertake Packaging

1.1 Receive treated seeds from seed treatment area

- Seed treatment area is normally situated between shed and ventilated storage building where a hall is connected to ventilated flat stores by covered gallery for easy movement of processed seeds.
- Seed treatments are chemical or biological substances that are applied to seeds to control disease organisms, insects or other pests.
- Seed treatment uses pesticides such as bactericides, fungicides and insecticides.
- Sacks, containers, trucks, wagons, augers, and conveyors are used for transporting treated seeds.
- Once the seed treatment is done, then handling of treated seeds should be done safely as per following instructions:
    - Comply with all label instruction for seeding operations.
    - Use high quality dust-free seed.
    - Use available engineering and system controls to minimize exposure to the seed treatment product and to ensure accurate application.
    - Use specified Personal Protective Equipment.
    - Wear dust masks when opening and/or emptying treated seed packaging to minimise contact with dust.
    - Ensure workers are adequately trained with training documentation.
    - Properly dispose of any spillage to prevent exposure to humans, animals, or the environment.
    - Ensure emergency contact information is easily accessible.
- After the treated seeds are handled as per instructions, the such seeds needs to be transported safely outside the seed treatment area by following:
- Treated seed should always be transported separately from people, animals and food and feed.
- Avoid undue abrasion and other mechanical damage to treated seed.

- Protect treated seed from heat and moisture. Shield seed and components from sunlight and extreme temperatures.
- Where treated seed is spilled, it should be collected immediately to reduce exposure risks.
- Read and understand seed treatment tag language and follow requirements that are printed on or attached to treated seed containers.

1.2 Receive treated seed test report from lab and check that all quality parameters of seeds as per SOP

1.2.1 Seed Testing lab

- Seed testing gives accruable and reproducible results regarding the quality status of seed samples submitted to Seed Testing Laboratories.
- Seed testing laboratories are organization in seed certification and seed quality control programmes which serve the producer, consumer and seed industry by giving information on seed quality.
- Test results of seeds can lead to rejection of poor seed multiplication or low grade seed in count of law.

1.2.2 Lab Seed Test Report Details

- The lab test report of treated seeds received present the results of studies and detailed description about test conditions, with thorough discussion of results, analysis of data, and conclusions drawn from analysis.
- Lab test report shows the summary and abstract about the seed test results which includes:

Test substance:

- chemical identification data, relevant properties of substance tested like physical state and stability;
- details on preparation of test solution

Test species:

- details of test organism: species/variety, plant families, scientific and common names, source and history of seed, viability, etc.
- number of mono-and di-cotyledon species tested;
- description of seed storage, treatment and maintenance.

Test conditions:

- testing facility i.e. growth chamber, phytotron, greenhouse;
- description of test system (e.g., pot dimensions, pot material, and amounts of soil);

- soil characteristics (texture or type of soil: soil particle distribution and classification, physical and chemical properties including % organic matter, % organic carbon, and pH);
- soil/substrate (i.e. soil, artificial soil, sand, others) preparation prior to test;
- description of nutrient medium if used;
- application of the test substance: description of method of application, description of equipment, exposure rates and volumes including chemical verification, description of calibration method, description of environmental conditions during application;
- growth conditions: light intensity, photoperiod, day/night temperatures, watering schedule and method, fertilization;
- number of seeds per pot; number of plants per dose; number of replicates (pots) per exposure rate;
- type and number of controls (negative and/or positive controls, solvent control if used)
- duration of the test.

Results:

- table of all endpoints for each replicate, test concentration and species;
- for the seedling emergence and growth test, the number and percent emergence as compared to controls;
- biomass measurements, i.e. shoot weight (fresh or dry) or shoot height of the plants as percentage of the controls;
- percent visual injury and qualitative description of visual injury (chlorosis, necrosis, wilting, leaf and stem deformation, as well as, any lack of effects) by the test substance as compared to control plants;
- a description of the rating scale used to judge visual injury, if visual rating is provided;
- for single rate studies, the % injury should be reported;
- EC_x/ER_x (e.g. EC_{50}/ER_{50} or EC_{25}/ER_{25}) values and related confidence limits for the endpoints and equation, calculated from the appropriate dose-response data and using the appropriate statistical procedures;
- description of the statistical procedures and assumptions used;
- graphical display of data and dose-response relationship.

1.2.3 Minimum Seed Standards

The probable minimum seed standards of respective seed crops will be shown in table:

Crop	Submitted sample (in gm)		Working sample	Phy. purity %	Germi-nation %	Moisture %	Other Distinguishing Variety	
	Certified Sample	Service Sample					FS	C
Cereals &Millets								
Paddy	400	50	40	98	80	13	0.05	0.2
Cumbu- Hy & var	150	25	15	98	75	12	0	0
Ragi	60	25	6	97	75	12	0	0
Sorghum- Hy & var	900	100	90	98	75	12	10/kg	20/kg
Maize - inbreds	1000	500	900	98	80	12	0	5/kg
Maize - others	1000	500	900	98	90	12	5/kg	10/kg
Barley	1000		120	98	85	12	10	20
Oats	1000		120	98	85	12	10	20
Wheat	1000		120	98	85	12	0	0
Pulses								
Redgram-Hy &var	1000	150	300	98	75	9	10	20
Black gram	1000	100	150	98	75	9	10	20
Green gram	1000	100	120	98	75	9	10	20
Cowpea	1000	150	400	98	75	9	5	10
Bengal gram	1000	400	1000	98	85	9	5	10
Horse gram	500	100	500	98	80	9	5	10
Peas	1000	250	900	98	75	9	5	10
French bean	1000	450	700	98	75	9	5	10
Field bean	1000	450	600	98	75	9	0	0

Crop	Submitted sample (in gm)		Working sample	Phy. purity %	Germi-nation %	Moisture %	Other Distinguishing Variety	
	Certified Sample	Service Sample					FS	C
Oil seeds								
Ground nut	1000	500	1000	96	70	9	0	0
Sunflower-V/Hy	1000/250	100	250/125	98	70	9	0	0
Gingelly	70	25	7	97	80	9	10	20
Castor	1000	250	1000	98	70	8	5	10
Soybean	1000	150	500	98	70	12	10	20
Fibre crops								
Cotton (Hy)	350-L 250-DL	200 100	35 25	98	65	10	0	0
cotton (var)	1000-L 350-DL	350 150	350 35	98	65	10	0	0
Vegetable crops								
Ash gourd	700	100	70	98	60	7	0	0
Bittergourd V/Hy	1000	250	450	98	60	7	5	10
Bottle gourd	700	100	70	98	60	7	0	0
Ridgegourd Hy/var	1000	150	400	98	60	7	5	10
Snakegourd	1000	250	250	98	60	7	0	0
Watermelon V/Hy	1000	100	250	98	60	7	5	10
Cucumber V/hy	150	100	70	98	60	7	0	0
Pumpkin V/Hy	350	100	180	98	60	7	0	0
Fruit vegetables								

Crop	Submitted sample (in gm)		Working sample	Phy. purity %	Germi-nation %	Moisture %	Other Distinguishing Variety	
	Certified Sample	Service Sample					FS	C
Brinjal V/Hy	150	10	15	98	70	8	0	0
Chillies	150	10	15	98	60	8	0	0
Bhendi	1000	100	140	99	65	10	10	20
Tomato V/Hy	7(hy),70	10	7	98	70	8	0	0
Greens	70	50	7	95	70	8	10	20
Cole crops								
Cabbage	100	10	10	98	70	7	0	0
Caulliflower	100	10	10	98	65	7	0	0
Knol kohl	100	10	10	98	70	7	0	0
Bulb crops								
Onion V/Hy	80	10	80	98	70	8	0	0
Root crops								
Carrot V/Hy	30	10	3	95	60	8	5	10
Beet root	500	50	50	96	60	9	0	0
Radish V/Hy	300	50	30	98	70	6	0	0
Turnip V/Hy	70	10	7	98	70	6	0	0
Cluster bean	1000	100	100	98	70	9	10	20
Forage crops								
Sunhemp	700	50	70	98	80		0	0
Daincha	900	25	90	98	70		0	0
Lucerene	50	25	5	98	80	10	0	0

Crop	Submitted sample (in gm)		Working sample	Phy. purity %	Germi-nation %	Moisture %	Other Distinguishing Variety	
	Certified Sample	Service Sample					FS	C
Cenchrus	25		3	80	30	10	0	0
Stylosnathus	70		7	90	40	10	10	20
Kolinji			90		70		0	0
Miscellaneous								
Palak	250	25	25	96	60	9	0	0
Spinach	250	25	25	98	70		0	0
Papaya	400		40	98	60	12	0	0
Moringa	1000		750	96	70	8	0	0
Coriander	400		40	95	65		0	0

1.3 Making arrangements of seeds storage materials like plastic bags, bins etc.

1.3.1 Seed storage

- Seed storage maintains the seed in good physical and physiological condition.
- Seed storage helps in enhancing the longevity of seed for sowing in following season.
- Pesticide-treated seed must be stored in dry, well ventilated location separate from untreated seed
- Some polyethylene or foil lined bags are also good containers for storing of treated seeds.
- The seed quality can be considerably be affected at any stages or segments mentioned above unless sound principles involved in seed storage are practiced and seeds are properly handled.
- Tested and certified seeds of cereals, pulses and oil seeds are normally packed either in gunny bags or cloth bags, however, paper bag, aluminium foil pouches and polyethylene bags are used for packing flower and vegetable seeds.
- The idea of treated seed storage is to preserve or maintain the physiological quality of seed for period desired by minimizing rate of deterioration.
- As seed storage is concerned with the control of deteriorative processes, so certain knowledge of such processes is required for successful seed storage operations.
- Cotton bags for seeds are made from sheeting, print cloth, drill, osnaburg and special seamless material.
- Cotton fabrics are produced in different widths and weights, each designed for special purpose.
- Cotton bags protect physical quality, but have no effect upon moisture content, insects, or rodents, unless special barrier properties are built in.
- The two most widely used laminated moisture proof bags consist of one or two sheets of paper attached to fabric with asphalt or compounded latex adhesive while other moisture proof constructions utilize such barrier materials as vegetable parchment.
- Most small packets are made of bleached sulfite or bleached kraft paper, which is coated with white clay to facilitate printing.
- A typical paper stock for small packets has a basic weight of 70 pounds per 500 sheets (25 X 38 inches).

- The packets are designed to contain a measured amount of seed without loss, but not to protect viability under unfavorable conditions.
- Many paper seed bags have multiple plies or layers which can be constructed of various thicknesses of smooth or crinkled paper.
- Some seed bags are made of laminates of paper/polyethylene/aluminum foil, so such combinations afford better protection against moisture than foil or polyethylene used alone with paper.
- Films of cellophane. Pliofilm, polyester, polyvinyl, aluminum foil, and polyethylene are used alone in various combinations.
- Cellophane which is made of regenerated cellulose is produced in more than 100 varieties which is designed for specific purposes.
- Moisture proof types having low moisture-vapor transmission rates, are used for small packages.

1.3.2 Seed package/container

Packaging of seed involves:

- Filling of seed bags to exact weight
- Placing leaflets in seed bags regarding improved cultivation practices
- Attaching labels with information related to kind of seed, variety, purity, germination percentage, date of germination test percent weed seed etc
- Storage/shipment of seed bag

Seed Container

- After drying the seeds to correct moisture content, seeds can be placed in sealed moisture proof seed package.
- Containers appropriate for seed storage include sacks, bags, baskets, tins, and plastic or glass jars.

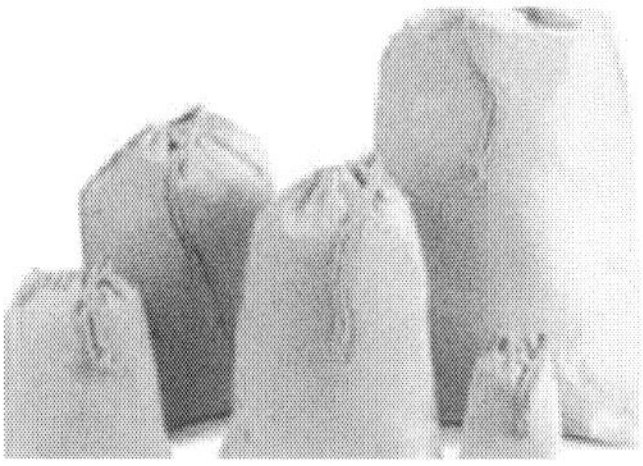

- Seed are packaged in containers having different sizes which starts from packets that can hold one gram of seed to bulk bins, that can hold tons of seed.
- Large quantities of seed can be stored in sacks, bags, baskets and tins.

- ❖ Small quantities of seed, high-value seed, and seed of species that produce seed infrequently should be stored in glass or plastic jars with air-tight lids.

In determining the kinds of container, the following points needs to be considered:

- ❖ quantity of seed desired in each package
- ❖ protection desired
- ❖ cost of package
- ❖ value of seed
- ❖ storage conditions in which, it is to be placed
- ❖ facilities for drying seeds

1.3.3 Types of Seed Packing Materials/Containers

Seed quality is affected by variation in moisture content, relative humidity and temperature, so to combat such factors, seed processing workers should store treated seeds in moisture vapour proof containers like polythene bag, aluminium foil, tin or any sealed container to maintain seed quality for longer time.

Gunny bags

- ❖ Gunny bags are available in different capacities which ranges from 35kg to 100 kg with or without inside plastic lining
- ❖ Such bags are mostly used for short term seed storage.
- ❖ Normally farmers prefer gunny bags for storage of farm cary over seeds.

Moisture and vapour pervious containers

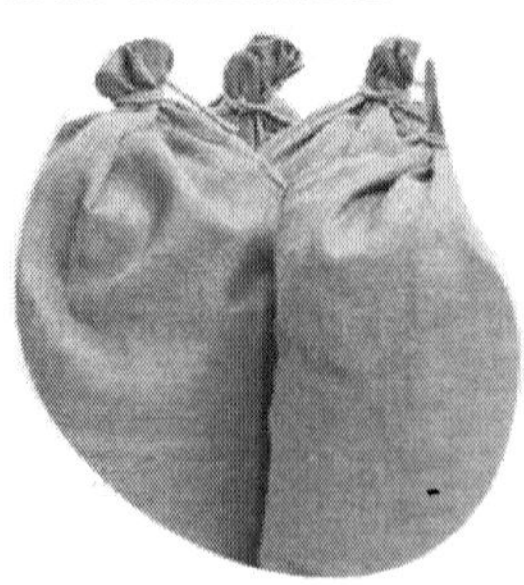

- The containers allow entry of water in from of vapour and liquid.
- Ii is good for short term storage.
- The seeds in these containers will attain seed equilibrium moisture with surrounding atmosphere.
- It includes cloth bags, gunny bags, paper bags etc.

Moisture impervious but vapour pervious containers

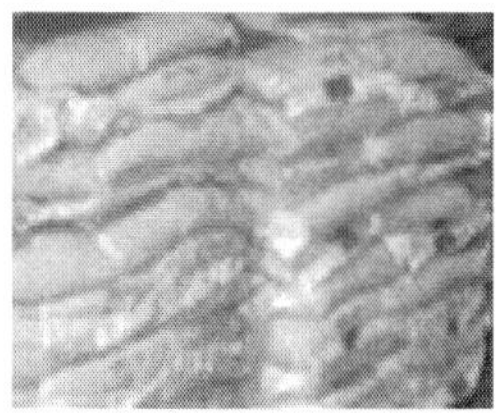

- The containers allow entry of water in form of vapour and not in liquid.
- The seeds in containers can't be carried over for long period in hot humid conditions.
- It includes polythene bags (<300 gauge thickness) and urea bags.

Moisture and vapour proof containers

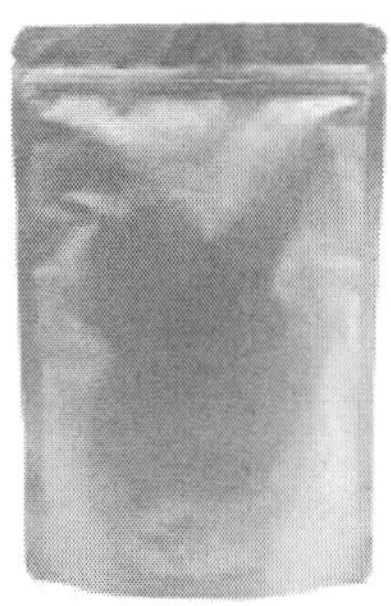

- The containers will not allow entry of moisture in form of liquid or vapour.
- It is used for long term storage even in hot humid conditions if seeds are sealed at optimum temperature.
- Polyethylene is not suitable for long-term storage of orthodox seeds for genetic conservation, because there is no absolute control over moisture uptake by the seeds.
- It includes polythylene bags (>700 gauge thickness), aluminium foil pouches, rigid plastics etc.

1.3.4 Precautions while storing treated seeds

- Read the seed treatment tag language to understand specific treated seed storage requirements.

- Treated seed should be protected from direct sunlight, extreme heat and moisture with facility to be well ventilated.
- Treated seed should be kept secure from children, livestock, wildlife and unauthorized persons.
- For storing treated seeds, make sure that such seeds are thoroughly dry before bagging, as excessive moisture can cause rapid deterioration of seed.
- Clearly label the bag of treated seed with proper indication of type of seed treatment.
- If the treated seeds need to be stored for year or more, in such case, seed worker should check the germination percentage prior to sale of such seeds.
- The treated seeds should never be stored in bulk storage bins that might also be used for edible grain storage.
- Store treated seed in special multiwall (3 or 4 ply) or tightly woven bags.

1.4 Weigh seeds for packaging to use suitable packaging materials as per company SOP

1.4.1 Determination of seed weight

The weight of the treated seeds can be determined by:

Automatic weighing and counting machine:

- In this, the treated seed sample is kept through the machine where number of seed in the sample may be noted.
- Weigh the sample to the same number of decimal places as in purity analysis.

Manual counting and weighing of seeds:

- In this, weight of the seed may be determined either on 1000 seeds singly or at least in 8 replicates of 100 seeds each.
- If the weight to be determined on 1000 seeds, then pure seeds should be counted manually from entire treated seed fraction of purity analysis and weighed in gms to similar number of decimal places as in case of purity analysis.
- For such purpose, at least 8 replicates of 100 treated seeds each from the sample should be counted by hand.
- In this, each replicates should be weighed in gms to similar number of decimal place as in purity analysis.
- If coefficient of variation does not exceed 6.0 for chaffy grass seed or 4.0 for other seeds, the result can be calculated.
- If the coefficient of variation exceeds these limits, count and weigh another 8 replicates and standard deviation should be calculated on sixteen replicates.

- If any replicate differs from mean by more than twice the standard deviation, it should be discarded.
- Calculate the average weight of 1000 seeds from weight of 100 seeds replicates

1.4.2 Weighing the working sample

The number of decimal places to which the working sample and the components of the working sample should be weighed is given below.

Weight of the working sample (g)	The number of decimal places required	Example
<1	4	0.7534
1-9.999	3	7.534
10-99.99	2	75.34
100-999.9	1	753.4
1000 or more	0	7534

1.4.3 Seed Bags as per weight

- The seed package/bag so selected by seed worker should form an association in mind of the farmer among seed and supplier with bag showing information about product related to its safety, storage and use of seed.
- Once the weight of seeds is determined, the seeds get packed in seed bags as per their weight.
- It is important that the pack is adapted to market requirements in terms of size, weight and unit cost.
- If seed worker uses over-large package sizes, then retailer will break bulk to sell the seed which provides opportunity for deterioration and adulteration which further invalidate the legal status of seed pack labelling.
- How seeds are packaged affects their physical characteristics of size; weight; color; moisture content; purity; and freedom from disease organisms, insects, rodents, and mechanical damage.
- Packages for treated seeds are made of burlap, cotton cloth, paper, films, metal, glass, fiberboard, and various combinations of materials.
- After the required seed quantity is filled in seed bag/container, then the seed bag should be weighed on a scale, so that net weight of seed with seed bag is known.

1.5 Select appropriate packaging materials like plastic bags, containers and bins etc. as per seed variety and as per company SOP

1.5.1 Introduction

- Seed packaging helps to maintain the quality while facilitating handling and identification of the product
- As quality of seeds needs to be maintained from time of packing to time of usage, seed workers needs to carefully select the packing material.
- The right packaging not only attracts consumers, it will keep the product safe in transportation and keep it fresh.
- The pack type selected will depend on nature of product and environment but packaging has to be resistant to moisture penetration, pest damage, adulteration, slipping in stack and to rough handling.
- Seed processing worker should always use standard sizes and shapes for product packaging, as this will improve flexibility and convenience during storage, handling and transportation, but also reduces production costs for bespoke packaging.

1.5.2 Different Packing Bags

- It is important to package seed in dry containers for proper storage.
- For small quantities of seed, processing worker should use containers such as tin cans, jars, or pots that are glazed on inside or reinforced boxes or bags.
- Metal or plastic cans, gunny bags or drums are often used to package large quantities of seed.
- Regardless of type of container used, the package needs to be of standard size and shape, so that if such material is filled with seed of known purity percentage, then an approximate number of seeds it contains can be estimated.

- The seed containers of standard sizes and shapes are easier to handle.
- For subsequent identification, each package of seed, or each aggregation of packages representing given seed collection, should be labelled.
- Cotton bags for seeds are made from sheeting, print cloth, drill, osnaburg and special seamless material.
- Cotton fabrics are produced in different widths and weights, each designed for special purpose.
- Cotton bags protect physical quality, but have no effect upon moisture content, insects, or rodents, unless special barrier properties are built in.
- The two most widely used laminated moisture proof bags consist of one or two sheets of paper attached to fabric with asphalt or compounded latex adhesive while other moisture proof constructions utilize such barrier materials as vegetable parchment.
- Most small packets are made of bleached sulfite or bleached kraft paper, which is coated with white clay to facilitate printing.
- A typical paper stock for small packets has a basic weight of 70 pounds per 500 sheets (25 X 38 inches).
- The packets are designed to contain a measured amount of seed without loss, but not to protect viability under unfavorable conditions.
- Many paper seed bags have multiple plies or layers which can be constructed of various thicknesses of smooth or crinkled paper.
- Some seed bags are made of laminates of paper/polyethylene/aluminum foil, so such combinations afford better protection against moisture than foil or polyethylene used alone with paper.
- Films of cellophane. Pliofilm, polyester, polyvinyl, aluminum foil, and polyethylene are used alone in various combinations.
- Cellophane which is made of regenerated cellulose is produced in more than 100 varieties which is designed for specific purposes.
- Moisture proof types having low moisture-vapor transmission rates, are used for small packages.

1.6 Pack seeds in the way to avoid any damage

1.6.1 Steps in Seed Packaging

- Decide most suitable seed bags/containers for storing seeds.
- Prepare and label the containers for each accession which can be computer generated self adhesive labels
- See that label contains information such as accession number, date of storage, weight of seeds, container lot, seed type, date of packaging

- weigh each empty labelled container before seed storage
- Fill the container with seeds and weigh again.
- add the label and seal the container to protect seeds from ambient RH
- Check the quality of each container after sealing it by visually inspecting the container to see any leakage
- Replace the container which is not fit/accurate
- Move filled containers in storage room

1.6.2 Packaging materials used for short-term seed storage

- Seed storage insects are very good at locating small openings in bags or containers that are not well sealed.
- There are many variety of materials that can be used to store seed for short-term storage which are non-rigid materials like cotton, burlap, paper and composite materials like multi-wall paper and plastic film, or polyethylene bags.
- Materials used for short-term storage are porous which adequately contain and protect seeds from mixing, and will not provide any protection from moisture or loss of seed viability.
- The materials for short term seed storage are used for mechanically separating seed lots, and for transporting and shipping seed till seed can be placed in environmentally controlled conditions for longer-term storage.
- Burlap bags have high strength, which can be re-used many times, and can be stacked high without slipping of stack.
- The strength of cotton bags depends on thickness of weave and thread and on quality of seams as it is not as strong as tear-resistant as burlap.
- Cotton bags can be re-used, depending on the quality of fabric.
- Bags made of woven plastic material are strong but tend to slip when piled high and are harder to close securely once bag gets opened.
- Multi-wall bags are made of several layers of paper in different types of construction with poor bursting strength if piled high or dropped, and if used repeatedly, can tend to become brittle along folds and wear points.
- Cardboard boxes and cans are expensive, but are re-usable, good for stacking which provides protection against mechanical injury to seed and to infestation by seed storage insects.
- Flexible packaging having weave such as burlap, cotton, or plastic offers little protection against seed storage pests such as grain moths.
- Seed stored in woven bags needs to be inspected if an outbreak is detected in particular seed bag.

- For small lots of seed, paper bags like lunch bags are inexpensive and adequate for storing seed, but seams are not always reliable, and if used, needs to be double or triple bagged in order to ensure integrity to prevent bursting.
- Package material like cellophane, acetate, and 2-4-mil polyethylene zip lock bags may be used.
- If using polyethylene, it is best to use 4-mil thickness, especially for heavy seeds.
- Plastics and thin films are not reliable moisture barriers, though such materials offer better moisture protection than paper.

1.6.3 Packaging materials used for long term seed storage

- Metal and glass containers are common to store seeds, so such containers needs to be properly sealed to prevent the exchange of moisture and gas as these are only reliable means of protecting seeds against humidity, insects, rodents, floods, and mechanical damage.

- Plastic bags should not be used for long-term seed storage.
- For storing large quantities of seed, metal five-gallon cans fitted with rubber gasketed lid and pressure ring are good for storing large seeds.
- One-gallon jars with gasketed seal lid are also good for storing large quantity of seeds.

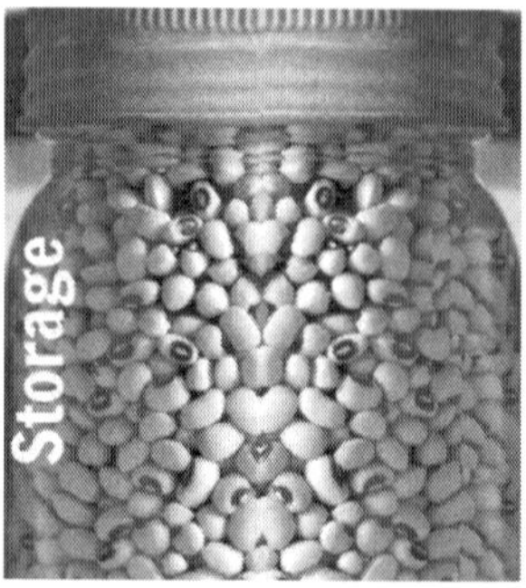

- ❖ One-gallon jars with plastic lids are flexible enough to form tight seal when lid is screwed tightly where metal lids requires rubber gasket.
- ❖ Gaskets can be cut from sheets of neoprene rubber or used automobile inner tubes, as glass is breakable, it has advantage of being transparent so that contents can be easily inspected for insect damage.
- ❖ Glass canning jars are available in different sizes which ranges from ¼ pint to ½ gallon.
- ❖ For storing small quantities of seed, ¼ and ½ pint jars are ideal, provided the rubber seal is in good condition.
- ❖ Other materials that can be used for storing seed can be of high-density polypropylene material with patented inner and outer valve to ensures airtight seal.
- ❖ Seed worker can use heat-sealable barrier pouch which are triple laminated material made from paper, plastic and foil.
- ❖ Heat sealable barrier pouch are sealed with heat sealer, or iron set on wool setting.
- ❖ If canning jars are used for storage, then seed worker should select particular brand name and style so that same size and same style containers have same weight.
- ❖ Once the weight of standard containers is known, it is easy to subtract that weight from container of seed to obtain weight of seed.

1.7 Pack seeds as per right quantity as recommended

- ❖ A package is a vehicle of safety and achieves objective of delivering safe, wholesome, nutritious value to consumer.
- ❖ Seed packaging involves placing a counted or weighted seeds in container which can be sealed for further storage.
- ❖ To prevent from moisture, seed should be packaged in waterproof containers and hermetically sealed.

- As per Indian standards, seeds can be packed in 100g, 200g, 500g, 1 kg, 2 kg, 5 kg containers, bags and thereafter in multiples of 5 kg bags.
- The number of seeds to be packed for storage depends on species being conserved and how often, seeds will be removed from monitoring, distribution or regeneration.
- As per the seeds standards, 3000 seeds are acceptable where material shows less morphological variation where 4000 seeds shows accession.
- The seed number can be converted from weight, if 100 or 1000 seeds weight are known.
- In declaring net quantity of treated seeds contained in a package, the weight of wrappers and materials other than seeds shall be excluded.
- As seeds are of different sizes and weight, so seed processing worker needs to correctly analyze the amount of seeds, which can be filled easily in seed containers, so that the quality of seeds can be maintained.
- Seed processing worker should take care that the required quantity of seeds to be filled in required bags/container.
- The bags or containers should not be loaded with extra seeds as it can cause tight sealing of bags/container which by pressure can be destroyed.
- Seed bags/containers/bins should not be overloaded with seeds as it allows uniform dissipation of heat and moisture in large airspace, so that warm moist air can be moved out of storage structure.

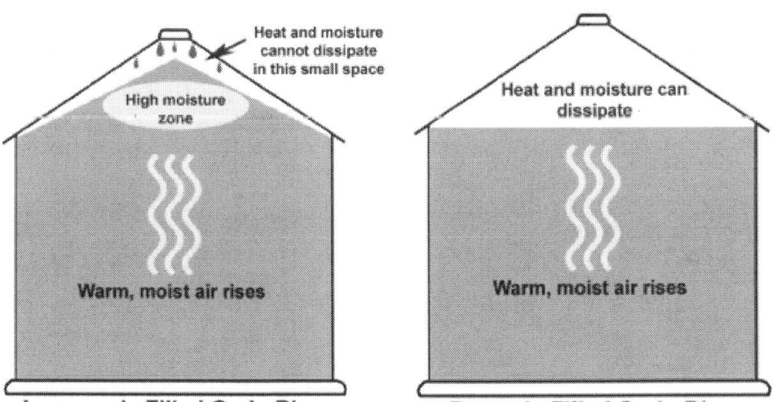

Improperly Filled Grain Bin | Properly Filled Grain Bin

1.8. Put required seal on packets, bags, containers and bins as per SOP

1.8.1 Label on Sealed Containers

- Before sealing the filled seed container, it is important for seed processing worker to put label on every container.

- ❖ Label is attached to seed bag/container, or can be printed directly.
- ❖ The label contains all-important information about the seed as complete processing records are required so as to trace the seed from time it is received till it is sold.
- ❖ Seed container should have relevant seed details which includes:

- ❖ Label No.
- ❖ Crop
- ❖ Variety
- ❖ Lot number
- ❖ Seed treated date, month and year
- ❖ Expiry date, month and year
- ❖ Germination percentage
- ❖ Physical purity
- ❖ Genetic purity
- ❖ Net weight
- ❖ Seed producer label
- ❖ Name and address of producer

Types of Labels

Tear - Resistant Labels

- Such labels are made of tear-resistant material and are tied or sewn to the container.

Adhesive Labels

- These labels must be affixed directly on container.
- Such labels have strong adhesive qualities and special cuts so that they cannot be removed without being destroyed.
- The labels should only be used on smooth or non woven materials.

Preprinted Labels

- The labels are printed directly on the bag with lot number.
- In this, additional lot information is printed on label during filling of bag.

Bulk Pedigreed Seed Tag

- The Bulk Pedigreed Seed Tag is official tag for bulk pedigreed seed.

1.8.2 Seed Container Sealing

- Once the seed container/bag is having a label, not it is important for seed processing worker to put a seal on such bags or container.
- To make the seed container/bag moisture-proof, sealing is required.
- The exact sealing technique will depend on the type of containers and sealing methods that are available.
- Laminated aluminium foil bags are easy to package, can be cut to size to save space and can be sealed again after us, but they are difficult to stack and must be made of good quality material and have good seals or leakage to occur.
- Seed cans are rigid and easy to stack and usually will not break open if dropped, but certain types of seed cans are not resealable, hence they are expensive to use.
- Glass bottles can be usually sealed again and amount of seed left is visible, but standard size bottles will waste the space and bottles can be break easily.
- Seeds can be packaged in bulk into large containers and/or sub-samples can be packaged separately into smaller containers.

1.8.3 Methods of Sealing Seed Containers/packets

Snap on Seals

- Snap on seals are made of metal which can be closed by squeezing between the finger and thumb to be closed with snap sound.

- Such seals are designed to break along the spine as evidence of tampering.
- As identifier, there exists logo, sign or sequential number printed on each seal.
- Seals are tamper-proof and must not be freely available to non-approved samplers/companies.
- The seal must be applied to string, tie or stitching so it cannot be removed without destroying integrity of seal.
- The official seed sampler may delegate attachment of seals to other individuals under seed supervisor.

Adhesive Patches, Labels and Tapes

- These are used to close the seed containers or holes which is made in containers.
- Such seal should have strong adhesive qualities and special cuts in label so that it cannot be removed without destroying integrity of label and therefore the seal.
- Such labels can close, seal and mark container in one step.
- The labels should have sequential numbers to identify the containers.
- such labels may not be effective in dusty conditions, on certain containers (cardboard boxes, woven plastic) and are only effective in specific temperature ranges.

Self-sealing Containers

- Self-sealing container has sleeve-shaped valve as part of bag.
- Once the bag is filled, the pressure of the seed closes the valve inside the bag.
- A direct sealing method must be used to seal the bags such as, sealing tape, a label, a patch, by gluing the mouth of the bag or sealing with a metal clip.

Sealing by Stitched Label

- In this method, sealing is done by sewing a single line of stitching through the label and closing the bag or adhesive labels, that needs to be glued directly on container.
- Care must be taken to verify that the stitching does not perforate any information on label.
- If seed lot needs issuance of OIC, and only company tags are stitched on containers, all of containers in seed lot should be uniquely identified by lot designation/number.
- The container should also be sealed using metal seals or non-removable adhesive labels.

Hermetically Sealed

- Containers which cannot be opened without causing damage to package may be regarded as sealed without undertaking any additional sealing.
- The packages can be paper bags or boxes that are closed with glue on all sides, sealed tins, sealed bottles or heat sealed foil packages.
- Such containers must have printing design on package which indicate if part of original container is missing.
- The containers are difficult to sample as they are destroyed in sampling.
- The repackaging of containers should be done in presence of sampler.

Tie / Strip Seals

- The tie/strip seals are made of plastic.
- Such seals are closed by drawing the end through self-locking device which saves from being opened without destroying the label.
- When such seal has unique identifier, it could become a seal.
- As a best practice, an inventory of the seals should be maintained.

Stapling of Labels on Stitched Bags

- This method of sealing may be used when the company has stitched the containers closed and then staples the labels onto the bags, provided that all the containers in the seed lot are uniquely identified.
- Such type of sealing method is not acceptable for sampling for issuance of an OIC or for seed lots.

1.9. Keeping record of packed seeds as per SOP

1.9.1 Records

Keeping track of packed seeds and knowing happening about packed seeds requires some records. Good field management requires good useful set of records which does not ensure successful of fields, but success is unlikely without them.

Record-keeping is a task that has been started many times by eager farmers/ growers, only to be dropped later. Reasons that complete, accurate records are not kept revolve around lack of time and motivation. Keeping records is productive work.

Good record-keeping system should be taken into account for seed processing workers as per needs, abilities and limitations. Seed workers should need to understand the purpose for keeping the records and when the purpose is known, it will help them in deciding which records are useful and need to be kept.

Records provide seed grower with data, information and knowledge. There are four uses for field records such as:

- ❖ Use as service tool which covers income tax calculations, estate planning, business arrangement reconciliation and obtaining and managing credit.
- ❖ Providing data for financial analysis and diagnostic instruments like identifying strengths and weaknesses of business.
- ❖ Progress indicator where series of records are required to monitor work progress.
- ❖ Forward planning tool for keeping past records to project future cash flows for current and modified farm management practices.

1.9.2 Details of Record

Records of field comprises of:

Field diary

- ❖ It is a form of small notebook where seed grower should record key facts and figures about farm business and its day-to-day activities.
- ❖ Farmers /seed grower who uses diary finds important facts and figures which could be easily lost or forgotten are stored permanently in this for future reference.

Unit returns

- ❖ It includes returns from all field outputs such as sale of crop, manure etc.

Crop stock accounting

- ❖ The main purpose of seed crop stock accounts is to monitor net losses and gains in income, and to distinguish the increases and decreases due to changes in market value.
- ❖ When quantifying total stock value, seed grower has to take into account both in change size and change in unit of the crop.

Key financial records

- ❖ It covers detail related to interest and principal repayment schedules to plan repayments.

1.9.3 Importance of Records

- ❖ They are management tool which allows seed workers to measure how efficiently they are using resources and to determine whether or not they are having income.
- ❖ The records are essential for planning and decision making
- ❖ The records are used in obtaining credit

- The records are used for income tax

1.9.4 Seed Packed Record details

Seed processing worker should keep relevant details about packed seed:

- Packet number
- Packing date
- Kind and variety of packed seeds
- Lot number
- Container type
- Packet Weight
- Quality of Packed seeds
- Name of packing agent

2

Undertake Labeling

2.1 Undertake labeling of packed seeds with self or computer generated self adhesive labels as per company SOP

2.1.1 Seed Label

- Before sealing the filled seed container, it is important for seed processing worker to put label on every container.
- Label is attached to seed bag/container, or can be printed directly.
- The label contains all-important information about the seed as complete processing records are required so as to trace the seed from time it is received till it is sold.
- Label provides end user with instructions on proper handling and storage of treated seed.
- Label ensures that treated seed is not used for any purpose other than for planting.
- Seed treatment pesticide labels needs extra information that seed treater should add to label of treated seed which includes special personal protection information, environmental hazards, statements of practical treatment in case of accident, or grazing restrictions.

As per seed law, seed treaters should print following information about treatment on same tag with analysis information, on separate tag attached to seed container, or directly on side or top of seed container:

- A word or statement indicating that the seed is treated, such as 'TREATED'
- The accepted common or chemical name of all pesticides applied
- A statement, as 'Do Not Use for Food, Feed, or Oil Purposes'

This Seed Treated With Captan Do not use for food, feed or oil purposes

- Where highly toxic pesticide shown by word 'Danger/Poison' on product label to be applied to seed with skull and crossbones and precautionary statement as 'Treated with Poison'.

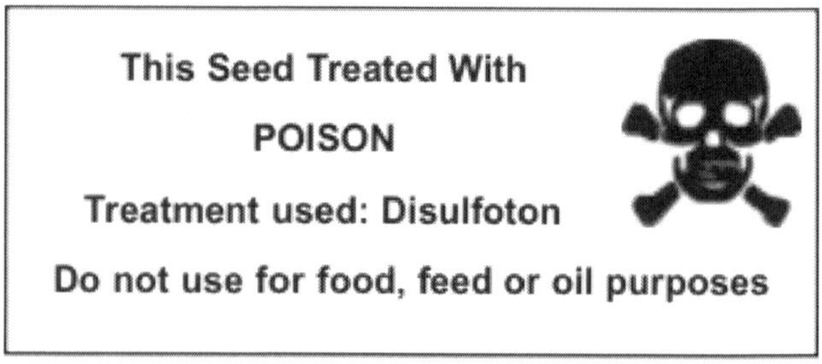

- Word 'poison' must be in type no under the hazard communication standards.

2.2 Check instructions on the label such as type of seed, quantity of seeds, season of harvest, date of packaging, expiry date, lot number, name of chemicals with which seeds were treated, caution statement, storage precautions and safe disposal of untreated seeds

Once the label is made and sticks using adhesives, it is the duty of seed processing worker to check for certain instructions on the label which should include:

- type of seed
- quantity of seeds
- season of harvest
- date of packaging
- expiry date
- lot number
- name of chemicals with which seeds were treated
- caution statement
- storage precautions
- safe disposal of untreated seeds

2.3 Check for any deficiencies and damages in seeds packaging material and sealing

In order to obtain optimum quality, seed processing worker should carefully examined the packaging materials as loose packaging will lead to moisture.

Checking of seed package is important for following reasons:

- ❖ To guarantee for quality of seeds after treatment, seed workers adopts many storage techniques, as qualitative losses takes place during this stage because of deterioration, which is stimulated by respiratory metabolic activity.
- ❖ The packed seeds due to hygroscopic, varies with fluctuations in relative humidity of air and temperature of environment, and experiences varying degree of humidity during period of storage.
- ❖ Such variations detrimental to maintenance of physiological quality, increasing deterioration and culminating in lipid peroxidation, disruption of cell membranes and disintegration of cell nucleus.
- ❖ To minimize losses during storage, seed workers adopts many technological alternatives for maintenance of seed quality, including use of impermeable packages, which tend to reduce environmental variations in seed metabolism.
- ❖ Use of carbon dioxide in packages reduces oxygen concentration and with increase in CO_2, metabolic rate of seeds reduces which makes viable alternative for conservation of seed quality.

2.4 Move the sealed packets for storage

2.4.1 Seed storage

- ❖ The aim of seed storage is to avoid germination cues and to slow the metabolism of seeds as much as possible so they live for longer.
- ❖ Generally seeds stored in moisture impervious containers like aluminum foil pouch, polythene bag, vacuum sealed containers etc. have storability for longer period compared to those stored in moisture pervious containers like cloth bag, paper bag, jute bag etc., under ambient storage condition.

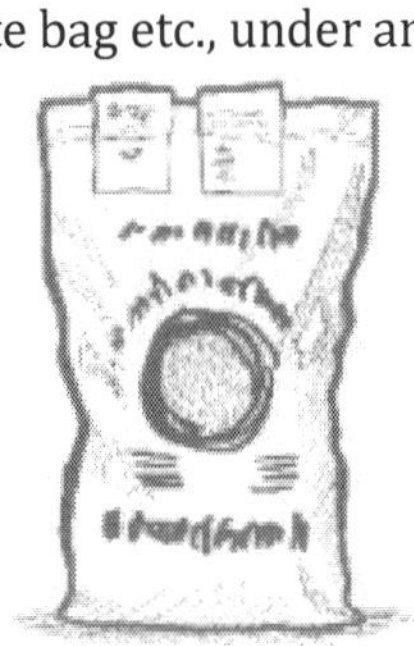

- ❖ The concept of storing seeds in moisture impervious sealed containers is to prevent mainly the migration of moisture content from the surrounding environment into the seeds.

- In sealed hermetic storage conditions, the seeds retain viability and vigour for longer period owing to lesser fluctuation of moisture content and temperature, decreased oxygen and enrichment of carbon dioxide in the containers

2.4.2 Moving sealed seed container to safe storage place

- Proper storage of seed is vital to conserving its vigor and vitality.
- Once the seed containers or bags are sealed properly and checked, seed worker now needs to placed the sealed containers in safe and proper place.
- Sealed seed containers/bags are transported and stored under pest free shelter, on pallets in locked up stores and away from walls to avoid damp before market.

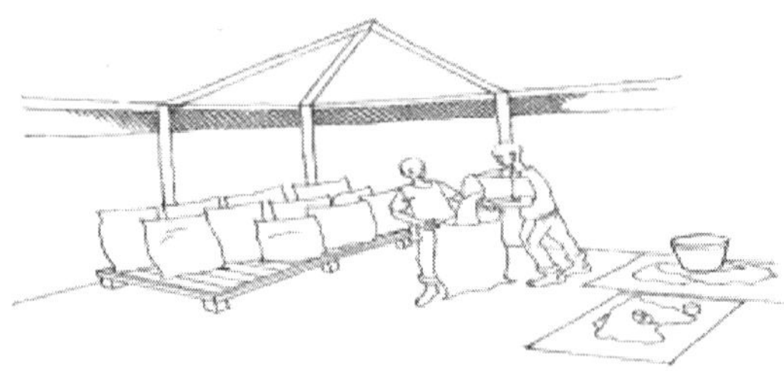

- If the packaged seed are lighter in weight, they can be transported manually or if the weight of the packaged seeds are more, they can be transported by gravity flow through pipes, airlift, belt conveyors or storage boxes or in bags.
- At very least, sealed seed bags should be kept off the ground to prevent spoilage by translocating water and/or termites.
- Seed contains/bags should be stacked on dunnage or waterproof sheeting, away from walls, in rodent proofed barn.

3

Achieve Productivity and Quality Standards

3.1 Ensure that seeds from different seasons are not mixed

3.1.1 Different seasoned seeds

- Many types of seeds can be collected which remains dormant till right conditions arise.
- Seeds come in different variety of shapes and sizes, with some like orchid seeds, being almost microscopic and others, like seed of palm tree which is around 12 inches long and weighing 18kg.
- The best time to gather seeds is when the weather is relatively dry and when the seed heads are ripe, making them easier to collect.
- Breeder seeds are produced in controlled environment, as such seeds have highest genetic purity level.
- Foundation seeds are direct result obtained by growing breeder seeds which are produced under supervision and control of breeder or releasing institution.
- Certified seeds are produced from breeder and foundation seeds having followed a set of approved procedures established by releasing institutions and or government agencies responsible for seed release.

3.1.2 Separating different seasoned seeds

- Once seeds of different seasons are collected in respective bags they are further checked by grading and sorting.
- In this, two sieve sizes with varying mesh sizes are used where one is little large than other.
- The two sieves are placed over each other where first allows anything smaller than desired seed size to pass through while other holds the seeds with larger size.

3.1.3 Checking of separated seeds

Once different seasoned seeds are separated of different sizes and variety, seed processing worker needs to endure himself, that the seeds seggrated, should be filled in respective containers.

3.2 Ensure that bags/containers/bins do not contain seeds more than their capacity

- Seeds during packaging must be done with care to avoid any possible effect on its physical purity.
- There are certain heavy seeds which gets fractured if they tends to strike against each other.
- Seeds while filling in seeds bags should be kept as per their weight and label information.
- For proper checking of required seed quantity in seed bags, seed processing worker should use beam scale or an elabourated scale to measure required seed quantity.
- If the seeds are flooded in seed container/bags, then there are chances that such bags can be tear off which lead to loss in seeds during transportation.
- Certain seeds are sold based on the number, either actual or adjusted for specific percentage, so each seed package should contain correct amount of seeds.

3.3 Ensure that label contain proper identification, instructions for maintaining the quality and traceability information

3.3.1 Label Identification

- Seed packages should be labelled properly to show species, variety, grade and lot number, percentage of live seeds, purity, content of noxious weeds, seed treatment.
- The label information should be printed on a tag attached to the container/bag which should be glued to the bag/container
- Seed label must include:

- Seed Variety and Kind
- Pure seed percentage by weight
- Percentage by weight of inert matter
- Percentage by weight of other agricultural seeds present
- Percentage by weight of all weed seeds present
- Percentage of germination
- Percentage of hard and dormant seed
- Total viable percent (germ plus hard plus dormant).
- Name and rate of occurrence of each kind of restricted weed seed present
- Lot number or other lot identification
- Name and address of person or firm who labeled the seed or who sells
- Seed origin
- Month and year germination test done
- Information about seed treatment

❖ If the seeds are treated, seed worker should see that there should be separate label, which should have information related to:

- Word or statement indicating that the seed has been treated.
- Commonly accepted coined chemical or abbreviated chemical name of any applied pesticide
- Description of process or commonly accepted name of substance applied if other than a pesticide
- If harmful to human or other vertebrate animals a caution statement such as "Do Not Use For Food or Feed or Oil Purposes"
- A poison statement or symbol for mercurial and similarly toxic substances

3.3.2 Seed Label with instructions for maintaining seed quality

❖ Seed quality plays an important role in the production of agronomic and horticultural crops.

❖ Seed processing worker should seed that the seed label should have information related to seed quality like:

- Trueness to variety
- Presence of inert matter
- Other seed crops
- Weed seed

- Germination percentage
- Vigor
- Appearance
- Disease free

3.3.3 Seed Traceability Information

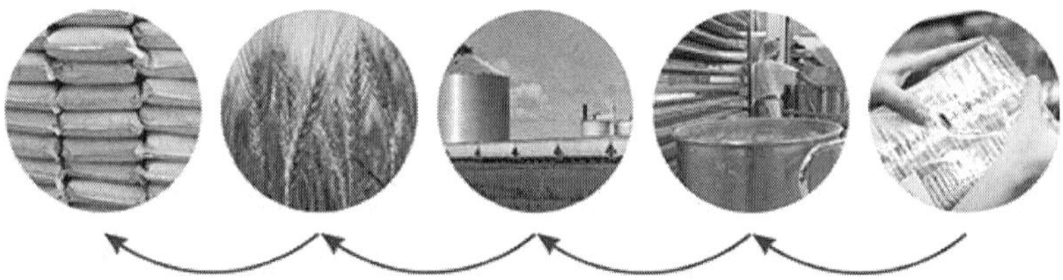

Traceability

- Traceability shows an ability to discern, identify and follow the movement of substance intended to be or expected to be incorporated in food by stages of production, processing and distribution.
- Traceability allows an organization to document or locate a product through the stages and operations involved in manufacture, processing, distribution and handling.
- Traceability creates confidence, trust and sustainability in trade relations for regulators, consumers, traders and buyers in supply chain.
- Seed traceability helps farmer to know from where the seeds have come, which can be done by using barcode or QR code.

- By tracing seed origins and testing of products along the supply chain, seed workers can helps in allaying of public's fears and can satisfy relevant legal requirements.

Factors affecting traceability

Major factors affecting the effectiveness of traceability are as follows:

- Supply chain structure and organization:

- Destination of a product;
- Identification of traceable lot unit;
- Time needed to trace a product;
- Credibility of traceability method;
- Data identification methods and data standardization
- Extent to which a traceability system is incorporated in an already existent and functional information management system and/or a quality/safety assurance system;
- Legislation on traceability

3.4 Maintain work area and equipment in clean and orderly condition

3.4.1 Cleanliness at work

Cleaning is a process of getting a clean, free, and energetic environment where there is no risk of health hazard. Clean surrounding makes it easier to organize the workplace so that it is easy to access all sort of material which are required.

Cleanliness is an important quality of life and serves as a habit that is often considered just next to godliness. It is a habit which is not related to wealth of a person, but which depicts the qualities of a person.

Cleanliness is nothing that should be forced on a person but should be properly practiced and be made a lifestyle. A clean and well-maintained person will always succeed in all difficulties if he/she applies it in every path he/she walks. It also helps to remain fit and healthy, mentally, physically and socially.

Cleanliness is required for health and wellness of an individuals and society. Health and cleanliness relates to each other and influences each other. In order to be in good health, one should practice hygiene. Hygiene is the practice of maintaining good health and preventing diseases through cleanliness. Cleanliness of body, food and environment contribute to the wellbeing of an individual health. It prevents disease as dirt usually dock on infected pathogens which causes disease to human beings when are exposed to them.

Cleanliness Rules

The importance of housekeeping at work can be clearly felt when you see the surrounding neat and clean as you enter the workplace. Though every company or organization comprises of housekeeping staff, it is not mandatory that they solely are responsible to maintain office cleanliness.

Employee too needs to understand the importance of cleanliness in the workplace and practice a few workplace hygiene tips so as to maintain a happy and clean office environment. Mentioned below are few workplace cleanliness policy rules that every employee need to follow.

Signs of poor workplace cleanliness

There are many signs which reflect poor cleanliness at workplace such as:

- cluttered and poorly arranged work areas
- untidy or dangerous storage of materials
- dusty and dirty floors and work surfaces
- items lying on work paths which are in excess or no longer needed
- blocked or cluttered work area or path
- tools and equipment left in work areas instead of stored properly
- broken pots, containers and damaged materials
- overflowing waste bins and containers
- spills and chemical leaks

Reasons for Cleaning and Hygiene Rules for workers:

- No cluttering of material used
- Always use gloves and proper hand sanitiser
- Make use of rubbish bin and proper damping policy
- Removing of unwanted plants, grass from lawn area
- Usage of proper working tools
- Keeping proper stock of chemicals and other materials properly
- Have regular discussion on health and safety matters
- Disposal of boxes and packaging covers should be done immediately

3.4.2 Cleanliness at workplace

A clean workplace can be said to as a place where there is all-round protection from germs to health. It is not a place which looks clean but is a place which is actually germs free for the workers and customers. A clean and organized workplace environment will not just be beneficial for worker but also will help in getting more customers. Cleanliness helps in increasing productivity as dirty place and clean mind cannot work together.

Workplaces in many years have realized the importance of clean and healthy surrounding. There are many reasons for which, one should maintain cleanliness and hygiene at workplace. As workplace is where workers spend several hours during their working, so it is advisable that it should remain clean, germs-free and healthy.

In workplace, the same concept applies as productivity is influenced by the environment. A clean environment will increase productivity as people will work with positive mindset that is free of stress. In health, cleanliness is important as hygienic practices contribute to good health of individuals. A dirty environment predisposes people to diseases, which affects their ability to perform activities and it results in less productivity.

3.4.3 Maintain seed machinery and equipment

The machineries used in seed processing involves:

- augers
- tubulators
- cup elevators
- bag sewers
- extractors
- air compressors
- vacuum cleaners
- ladders

Seed Processing worker while seed maintaining machinery and equipment should see that:

- Minor repairs are done without risk of injury to self or others and equipment or machinery should be returned to service with minimum down time as per company procedures.
- Preventative maintenance is carried out as per manufacturers' recommendations and company procedures.
- Preventative maintenance is carried out as per equipment maintenance schedule.
- Minor repairs are carried out on machinery and equipment and are recorded as per company procedures.

3.4.4 Maintain seed store area

Seed Processing worker while maintaining seed store area should see that:

- Blocks and alleyways are clearly identified as per manual and company procedures.
- Containers of dressed seed are free from dust and surface seed and are in condition as per manual and company procedures.
- Presence of vermin and storage pests is identified, and action is taken to control them as per manual and company procedures.
- Offal is segregated from machine area and kept away from seed as per manual and company procedures.
- Seedlots are kept as per manual and company procedures.
- Floors around the processing area are kept clean as per manual and company procedures.

3.5 Ensure proper documentation as per company's SOP

Documents which are part of seed processing/packaging system needs to be documented properly based on complexity of task, method used and skills.

Documents refer to detail which describes:

- System/Department/Non Technical Procedures
- Quality policies
- Technical procedures
- Standard operating procedures
- forms
- References

Proper documentation of seed packaging/storage procedures allows rapid accession of new samples; response to queries regarding conserved germplasm; and monitoring of quality and quantity of stored material to carry out regeneration and distribution which includes:

- Storage conditions/collection type
- Type of container, if this varies in the genebank
- Number of containers
- Total amount of seeds stored (by weight or number)
- Date of storage
- Location in genebank
- Minimum amount of seed allowed (base unit) for dissemination/ regeneration
- Location of safety duplicate, if available

Unit 4: Store Seeds

1

Prepare Seed Storage Room/Godown

1.1. Ensure cleanliness of environment around storage room / godown as per SOP

1.1.1 Seed Storage

The major steps involved in seed storage are:

- ❖ Dry and cool storage conditions.
- ❖ Protection from pests.
- ❖ Sanitation of seed storage area.
- ❖ Drying of seeds before storing.
- ❖ Storing well cleaned seeds in the storage godown.

1.1.2 Clean Workplace

- ❖ Cleaning is a process of getting a clean, free, and energetic environment where there is no risk of health hazard.
- ❖ Clean surrounding makes it easier to organize the workplace so that it is easy to access all sort of material which are required.

- Cleanliness is an important quality of life and serves as a habit that is often considered just next to godliness.
- It is a habit which is not related to wealth of a person, but which depicts the qualities of a person.
- To maintain high standard of godowns cleanliness, proper cleaning of storehouses should be done and rubbish should be wiped out.

- To clean the workplace surrounding, it is the duty of workplace manager or seed processing worker, to check that workplace should be cleaned and the working environment around the workplace should not carry any unwanted or dump material.

The surrounding area shall be maintained by taking appropriate measures to protect the storage area from any potential contamination if any polluting industry is located nearby and also taking following actions on regular basis:

- land, roads, yards, parking lots outside the godowns shall be free of debris and refuse and from any source of pollution and that areas should not constitute a source of contamination due to dust or any stagnation of water.
- removal of litter, waste, weeds and grass within the immediate vicinity of buildings or structures, as such growth accumulation may constitute an attractant, breeding place, or harbourage for pests.
- Cleaning of dust debris, sweeping shall ensure that fine dust and contaminants are not contaminating the raw grains or processed products in any ways
- Drainages are devoid of any stagnation of water to prevent contamination.
- Precautions shall be taken to prevent contamination from trucks, vehicles, forklifts or by foot.

1.1.3. Unorganized and messy environment

- The messy environment is the main cause of all problems.
- In this, there are more chances of threats going unnoticeable and the situation may simply turn the place into accident-prone area.
- An occupational hazard is a result of ignorance on part of workers or carelessness on part of workers.
- If area around workplace is messy and unorganized, it may be a door to many uninvited hazardous accidents, so seed worker should properly arrange and clean the workplace surrounding areas.

1.2. Ensure cleanliness of storage room / godown as per SOP

1.2.2 Cleanliness of godowns

- Good hygiene in workplace and godowns is important in maintaining grain and seed quality, so seed processing worker should see that storage areas should be clean which includes sweeping the floor, removing cobwebs and dust, collecting and removing any grain spills.
- Seed workers should see that storage area should be clean once they are emptied by spraying on walls, crevices and wooden pallets using insecticide before using them again.

- Normally, godowns are located away from the city where you will find dust, so seed processing worker should see that the godowns or workplace should be free from dust.
- Clean workplace can be said to as a place where there is all-round protection from germs to health.
- Clean workplace looks clean but is a place which is actually germs free for the workers and customers.
- Storage areas should be kept clean and free from all refuse and incorrectly stored materials.
- A clean and organized workplace environment will not just be beneficial for worker but also will help in getting more customers.
- Cleanliness helps in increasing productivity as dirty place and clean mind cannot work together.
- Workplaces in many years have realized the importance of clean and healthy surrounding.
- There are many reasons for which, one should maintain cleanliness and hygiene at workplace.
- In workplace, the same concept applies as productivity is influenced by the environment.
- A clean environment will increase productivity as people will work with positive mindset that is free of stress.
- In health, cleanliness is important as hygienic practices contribute to good health of individuals. A dirty environment predisposes people to diseases, which affects their ability to perform activities and it results in less productivity.

1.2.3 Signs of poor workplace/godown cleanliness

There are many signs which reflect poor cleanliness at workplace/godowns such as:

- cluttered and poorly arranged work areas

- untidy or dangerous storage of materials
- dusty and dirty floors and work surfaces
- items lying on work paths which are in excess or no longer needed
- blocked or cluttered work area or path
- tools and equipment left in work areas instead of stored properly
- broken pots, containers and damaged materials
- overflowing waste bins and containers
- spills and chemical leaks

As workplace is where workers spend several hours during their working, so it is the duty of seed processing worker to see that the workplace/godown should remain clean, germs-free and healthy.

1.3. Ensure that storage room is free from insects, fungi and rodents

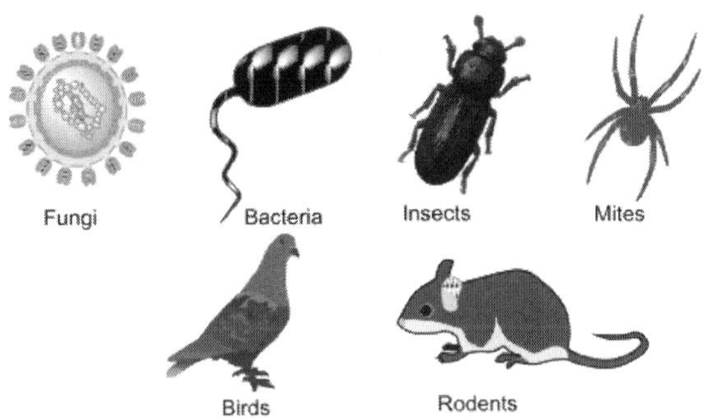

1.3.1 Seed storage pests

- Timely detection of storage pests plays a major role in seed control measures.
- Seed storage pests include insects, pathogens, rodents and birds.
- Such pests cause losses through a combination of feeding, spoiling and contamination of both paddy and milled grain.

Insects

- Each insect species has its own optimum temperature and moisture range conditions for development as well as a preferred feeding habit.

- The optimum temperatures for most insect species range between 25 and 32°C.
- At temperatures below 14°C and above 42°C, rate of development is reduced and most storage pests will die at temperatures below 5°C and above 45°C.
- The optimum relative humidity for most species is around 70%, with the minimum 25 - 40% and the maximum 80 - 100%. Very few species are able to survive in extremely dry conditions.
- Under good conditions, the duration of the development cycle from egg to adult is 18 - 25 days for beetles and 28 - 35 days for moths. In unfavorable conditions, this period may be extended to several months.
- Insect activity and fertility are affected by change in light.
- Moths are most active at dawn and at dusk.
- Inspections to check and control flying insects are best made at these times.
- Artificial light helps in reducing the movement and fertility of moths.
- Most storage pests are able to penetrate in seed and in seed bags more quickly and thoroughly than bulk grain due to gaps between the bags.

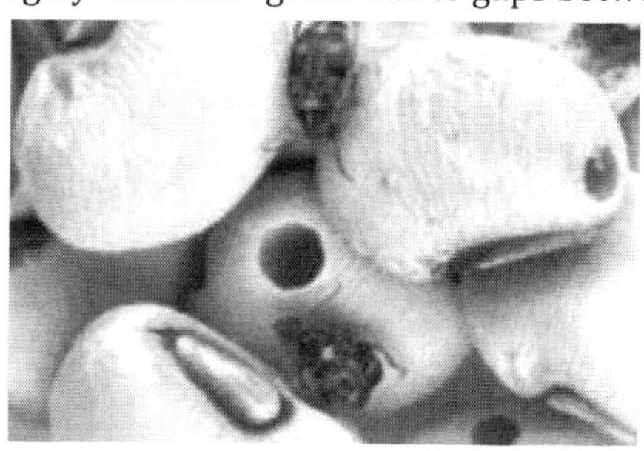

- The size, surface texture and nutrients in seeds also influence the ability of pest to attack the seeds.

Management of storage insects

- The management of stored seeds pests should be done in sequential and integrated manner.
- An effective pest control system will involve in:
  - Harvesting, drying and storage of clean dry grain
  - Disinfecting the storage system and
  - Controlling or preventing pest infestation during the storage period.

Fungi in stored seeds

- Contamination of seed with fungal organisms result in poor germination, seedling vigor or quality.

- Storage fungi invade seeds during storage and are not present in large quantities before harvest in field.
- The most common storage fungi are species of Aspergillus and Penicillium which are widely distributed and almost always present.
- Contamination occurs through small quantities of spores contaminating grain as it is going in storage, in handling and storage equipment or from spores already in storage structures.
- Under high temperatures and moisture, such small amount of inoculum increases rapidly.
- The development of fungi is influenced by:
  - Moisture content of the stored grain
  - Temperature
  - Condition of the grain going into storage
  - Length of time the is grain stored and
  - Amount of insect and mite activity in the grain.

Safe Storage Conditions

- Grain damage by fungi will be reduced when seed is:
- Stored at moisture contents below 13-14% as it is important to be aware that there is variation in moisture content through seed mass and fungi will grow where moisture is suitable and not as per average moisture content of seed stack.
- Stored at temperatures below 20oC and above 40ooC.
- Not cracked and broken kernels or contain large amount of foreign material as broken or cracked kernels are more likely to be contaminated going into storage and more likely to be invaded once they are in storage than whole kernels.
- Free from fungi coming into store as seeds moderately invaded by storage fungi can develop damage at lower moisture content, at lower temperature and in short time period than seed free or almost free of storage fungi.
- Stored for shorter period as seeds that is to be stored for only few weeks before processing can be stored safely with higher moisture content and more extensive invasion by storage fungi and can be kept at a higher temperature than grain that is to be stored for months or years.
- Free from insect and mites as insects and mites can carry fungal spores on their bodies thus introducing storage fungi in seed mass.
- Insect activity in a grain mass leads to increase in temperature and moisture content of seeds surrounding the insect infestation where hot spots conditions is favorable for mold growth.

Rodents in seed storage

- Rats have been estimated to damage more than 1% of world seed crops, where as in developing countries, it estimates to 3-5%.
- There are around 50 diseases which can be transferred to humans by rodents, including typhoid, paratyphoid, and scabies.
- Rodents may be vectors of large number of diseases affecting domestic animals.
- As rodents prefer food rich in proteins and vitamins and feed mainly on embryo, so they cause particular damage to nutritional value and germination ability of seeds.
- The three most important rodent species are:
  - Black rat or House rat
  - Norway rat or Common rat
  - House mouse

Rats and mice cause losses in a number of ways:

- Feeding on stored produce
  - Rats consume about 25 gm of seed in a day and mice eats around 3-4 gm in a day.
  - Besides eating stored seeds, rodents contaminate a lot of stored produce with urine, feces, hair and pathogenic agents.
  - As it is extremely difficult to remove contamination, infested batches often have to be declared unfit for human consumption.
- Damage to material and equipment
  - Produce leaking out of damaged bags or storage containers
  - Bag stacks collapsing due to damage to lower layers
  - Short circuits leading to sparks or fire from cables being chewed
  - Silos and warehouses may subside or even collapse as a result of being undermined
  - Drainage canals around a store may be damaged.

Preventive measures against Rodents

- The most essential factors for occurrence of rodents are:
  - Sufficient supplies of food
  - Protected places in which to build burrows and nests
  - Hiding places
  - Access to produce

- Good store management and preventive measures taken as part of an integrated control program can help to deal with such factors which includes:
    - Keeping godowns absolutely clean by removing any spilt grain immediately as it will attracts the rodents.
    - Store bags in tidy stacks set up on pallets, and ensures that there is a space of 1m all round the stack.

- Store any empty or old bags and fumigation sheets on pallets, and if possible in separate store area.
- Keep godowns free of rubbish in order not to provide the animals with any places to hide or nest.
- Keeping godown surrounding free from tall weeds so as not to give the animals any cover.
- Keeping godowns area free from stagnant water and ensuring rainwater to be drained away, as it can be used as source of drinking water for rodents.

1.3.2 Using disinfectants in workplace

Seed processing worker should see that the godowns and workplace should be disinfect with proper pesticides, so that the godowns should be free from rodents, fungus and insects.

- Disinfectants are very important to be used in work area, passage, godown as there are more risk of insects, fungus and rodents.
- The disinfectants comparatively lessen the chance of infection which can be through insects, pests, rodents etc.

- Chemicals such as Phosphine, Methyl Bromide, Dichlorvos, Carbondioxide, Ethylene oxide and HCN should be used in godowns for insects, fungus and rodents.

1.3.3 Treatment of premises

- Before any application of insecticide in seed storage area, entire area must be thoroughly cleaned.
- The range of available insecticide products is broad enough to treat different surfaces according to their characteristics.
- Treatment of premises should be given as:
  - Uneven surfaces should be treated by spraying with powder mixed with water.
  - Smooth, non-porous surfaces should be sprayed with stickier concentrate.
  - Ambient treatment should be used to destroy flying insects using aerosols in hermetically sealed premises, as such treatment is preferred in the evening, when flying insects are most active.

1.4. Ensure that storage room is dry, cool and well ventilated

1.4.1 Dry and Cool Seed Room

- Seed godowns should be dried and cool before storing seeds bags/bin.
- The seed must be dried before storage in godowns and store houses must be kept dry and well aerated before use.
- Seeds stored in cool, dry conditions will survive longer than seeds stored in wet, warm environment.
- It is important to leave adequate space between stacks of seed bags/ containers inside the storage room for air to move quickly.
- For seed room to be cooled, ensure that ventilation fan and air way in the room should not be blocked.

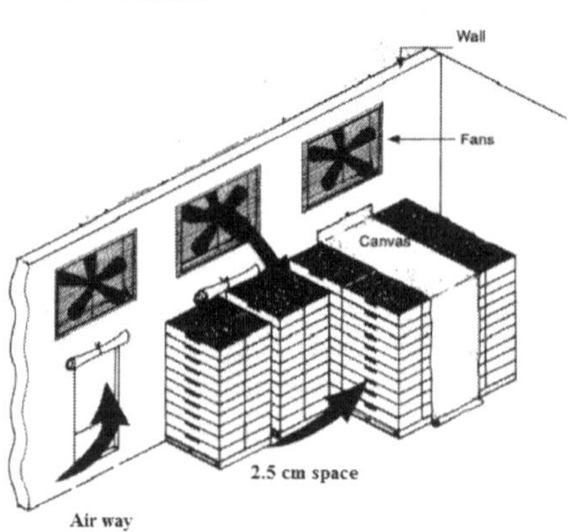

1.4.2 Ventilation in storage room

Ventilation

- Ventilation controls an indoor air environment with air flow.
- Godowns with proper ventilation will help in cooling an indoor environment as it gets very hot due to high concentration and type of materials stored in it.

- Ventilation openings are necessary for allowing the renewal of air and reducing the temperature in godowns as they allow some light to enter it.
- In godowns, ventilation is used to reduce seed temperature and seed moisture content.
- Ventilation will helps in preventing hot spots from developing, formation of convection air current and maintenance of uniform seed moisture content and temperature.
- Whether outside temperature of air and relative humidity are low enough to benefit the seeds, either by reducing seed temperature or moisture content in seeds, ventilating fans can be turned ON.

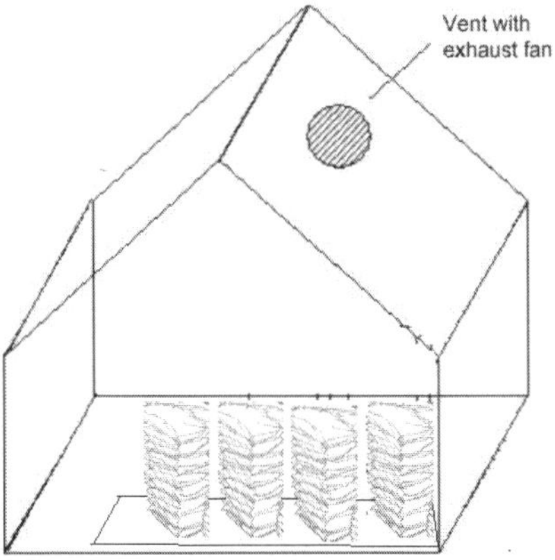

- If openings in godowns are located too low down they can be source of entry of water, rodents, thieves, etc, so to avoid this, ventilators are placed under the eaves.
- Ventilators should be fitted on outside with anti-bird grills and on inside with 1 mm mesh screens which will deter most insects.
- For proper ventilation in storage godown, vents should be located at ground level which can be opened at night, where fans can be used to pull cool air through the store area.

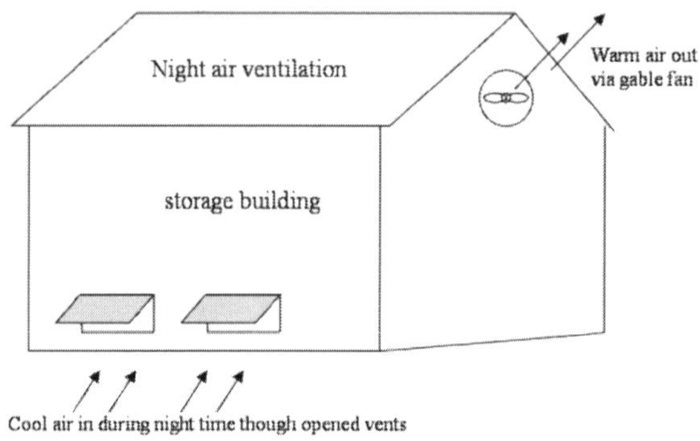

Ventilation in seed store room can be improved if air inlets are located at bottom of godown while air outlets are at the top.

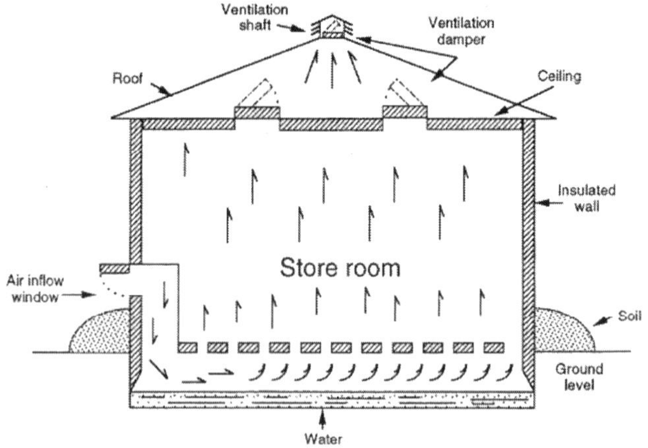

Ventilation Precautions:

- The moisture content of seeds should not be allowed to increase to a value in equilibrium with air relative humidity above 65%.
- The seed temperature should not increase above 33°C for more than few hours.

❖ Ventilators and doors should be opened during dry weather and closed during the period of excessive humidity.

1.5. Ensure that there is no leakage of rain water in the facility

1.5.1 Introduction

❖ Storage godowns should be waterproof to protect grain from moisture.

1.5.2 Causes of rain water leakage

❖ Seed worker should be extra vigilant when it comes to protecting seed bags/containers against expensive water damage.

❖ Incoming water is more than nuisance as it can be equivalent to huge repairs bill or adding damage slowly over time.

The rain water problem in godowns can be due to:

Mold and mildew

❖ Unseen water or high humidity can result in mold and mildew.

❖ Neither is as innocuous, but in fact, both pose safety hazards, and removal can be incredibly expensive.

Cracks

❖ A godown is prone to damage that can be made worse by presence of water through rains.

❖ Tiny cracks in walls, ceilings and floors allow moisture to enter in godowns.

❖ As temperature changes, moisture can be wrecking ball for structure as it expands and contracts.

Large-scale damages

- ❖ A burst pipe or extreme weather can wreak havoc on godown in short amount of time.
- ❖ It can do structural damage and destroy seed bags/containers.

1.5.3 Precautions

- ❖ The godown should have roof which should overhang the gables by 0.7 to 1.0 metres, and eaves by at least 1 metre to ensure that rainwater is shed well clear of walls.
- ❖ The overhang also helps to keep godown walls cool and protects ventilation openings from rain.

1.6. Ensure adequate lighting and drainage facilities

1.6.1 Lighting at Godown

- ❖ Adequate light in godown/warehouse is an important factor as far as safety of workers inside it is concerned.
- ❖ Sufficient and suitable lighting, whether natural or artificial, must be provided in every part of an undertaking where workers are employed or pass which includes outside yards and storage areas.
- ❖ Lighting in outside yards and storage areas should be such that dark shadows or glare is avoided on working area of stacks and storage fixtures.
- ❖ Lighting can be supplemented by painting appropriate surfaces white or off-white.
- ❖ Dirty light fixtures are indeed one reason to inefficient light supply, so seed worker should ensure that light fixtures in godown should be free from dirt, so that adequate lighting should be there.
- ❖ Adequate lights are required in order to do daily task as no proper lightening can be an unsafe combination with workplace.
- ❖ Work area, pathways, seed store area and seed equipment areas should be well lighted to prevent accidents and in contributing healthy and safe environment.

1.6.2 Drainage in Godown

- Water drainage system should be proper in godown.
- Seed/godown workers should see that there is no leakage in walls, otherwise flow of water in godown will generate/increase moisture in seed.

1.7. Ensure adequate stacking facilities like poles, wooden platforms etc.

1.7.1 Stacking

- Seeds have to be stored and preserved on scientific lines in godown till they are issued to growers.
- The bags containing seeds cannot be just dumped inside the godown, for it will not facilitate proper storage.
- Stacking is done so that seed/commodities can be used at later date.
- Good stacking make sure that seed will remain as close to its original condition as possible
- Seeds must be laid out and stacked in a way that makes sense to storage facility operators and their staff.
- Space for emergency and normal stacking must be set aside, but all stacks must be constructed as if they are permanent stacks.
- Floors or surfaces required to support stacks, shelving, racks or other means of storage should be capable of sustaining the intended load together with shock loads.
- Proper stacking ensures free access to the stocks in all parts of the godown for inspection and helps in effective disinfestations work.
- Wooden pallets should be used for arranging bags in cris-cross manner for effective ventilation on all sides of the bags.
- Seed bags should be stacked upto 6-8 tires depending upon density of seeds
- Size of stack should be 30x20 feet facilitate fumigation under gas proof or polythene covers.

1.7.2 Wooden Platform

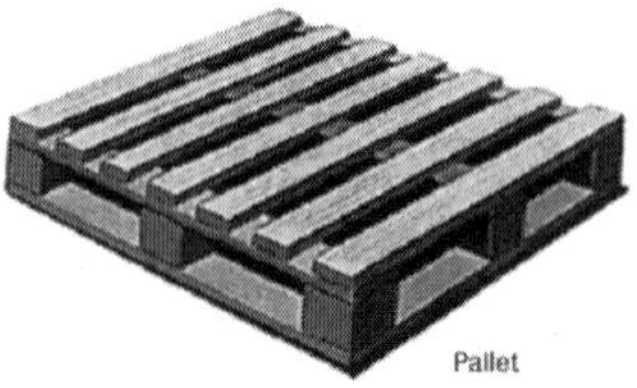

- Pallets should be of sound construction, and be of adequate strength for the loads and conditions under which they are used.
- Where pallet loads are stacked tier on tier, the lower pallets should be of suitable strength and in good condition and the unit loads must be able to support the weight above.
- The stability of stacked pallets or unit loads should be maintained by suitable bonding, avoiding excessive stack heights, to ensure that the contents of any pallet or unit load cannot collapse.
- The stability and structural strength of each pallet or unit load should be assured by bonding, taping, shrink wrapping or other means.
- When pallet or unit loads of cartons or sacks are stacked, care should be taken that they are not damaged by equipment.
- Pallets or other supports used for forming unit loads should be regularly inspected for damage and wear.
- Items which could cause damage to materials should be taken out of use until repaired, or be destroyed.

1.7.3 Positioning of Stacks

- Stacks should not be built within 450 mm of a wall. Most building and party walls are not retaining walls and may not withstand the pressure set up by stacks laid against them.
- A 450 mm gap between the stack and the wall also enables the construction and the condition of stacks to be checked.
- Stacks should not be created where any part will be within 1.5 m of a rail track.
- Extra care should be taken if the storage area is subject to vibration from rail or road traffic, outside or inside the premises.
- Care should also be taken in stacking material in the vicinity of machinery or operations which cause vibrations.
- Stacks, shelving and other fixtures for holding or storing materials should be so laid out and designed that there is sufficient access for safe loading and unloading by either manual or mechanical means.

- Stacks should have clear spaces of at least 1 m on all sides, apart from walls, where a 450 mm space should be allowed.
- Goods must not be stacked within 1 m of the ceiling, roof or sprinkler heads.
- Goods should be kept well clear of light fittings, heating pipes and appliances, firefighting and alarm equipment, and doors.
- No stack, shelving, fixture or other means of storage should be placed in a position, or extended in height, so that a person climbing on to it or removing stored goods, either manually or mechanically, can come into contact with live electrical wiring or unfenced machinery.
- No material should be loaded or unloaded from stacks, shelving or fixtures if there is a risk of workers directly or indirectly contacting unfenced machinery, or touching live electrical wiring, until that machinery or wiring is isolated or made safe.

1.7.4 Size and Shape of Stacks

- The size and shape of a stack depends on the storage space available and on the size, shape, bulk, weight, rigidity or fragility of the articles to be stored.
- The following are the commonest forms of stack:
  - Column—Single articles placed one above the other.
  - Square—Any stack, other than a column, with all sides vertical.
  - Pyramid—A stack in which the plan area is reduced in every succeeding tier.

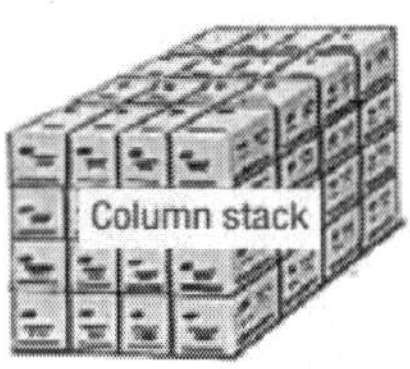

  - Stepped—Stacks with two or more adjacent tiers of the same area and each succeeding group of tiers of a smaller area than the group on which it is set.
  - Triangular—Stacks in pyramid or stepped form on two opposite sides, the other two sides being vertical.

1.7.5 Stack Inspection

- Racks, shelving, fixtures and other equipment used for storing, loading and unloading materials should be regularly inspected for damage and other defects which might cause loss of strength or result in damage or injury.
- Any damaged or defective equipment should be taken out of service until repaired.

1.8. Ensure security and fire fighting arrangements

1.8.1 Security

- Seed warehouse/godown should be surrounded by perimeter fencing or walls of a suitable height to ensure security of storage area against vandalism, theft or arson.
- Godown should have security gatehouse to monitor vehicles entering and leaving the godown/site.
- In godowns, there should be perimeter lighting and monitoring by closed-circuit television (CCTV).
- Access roads should be provided to all godowns on site so that vehicles can deliver goods to storage facility.
- Windows and doors at ground floor level are most vulnerable, so they should be fitted with good quality locks.
- Shutters or fixed security bars are also recommended on windows.
- Preferrably godowns should be protected by automatic intruder alarm system.
- Godowns should be kept locked and all keys should be kept under close control and should be distributed to nominated key holders only and regular checks should be made to ensure that they have not been lost.
- In godowns, fob and/or number code entry systems should be used instead of keys as such type of entry system eliminates the need for additional keys to be cut and allows access to be recorded on a computer.

1.8.2 Fire Fighting Arrangement

Fire-fighting equipment

Introduction

- ❖ Fire instruction notices should be prominently displayed in all storage areas.
- ❖ Godowns require adequate access to minimum percentage of work area perimeter so that fire and rescue service vehicles can reach the source of fire.
- ❖ Fire protection equipment such as automatic fire detection and alarm system needs to be designed, installed and maintained in all seed storage areas.
- ❖ Godown should have first aid fire fighting equipment such as hand held fire extinguishers as per relevant standards.
- ❖ To meet the requirements of godown regulations, manual alarm system should be minimum standard as required in warehouses, since there is no sleeping risk.
- ❖ For suppressing of fire, godowns should have automatic sprinkler system and smoke ventilation system as per company standards.
- ❖ Fires have occurred when the sacks are placed in store while smouldering.
- ❖ Where seeds are stored in pallet racks, there should be a provision of in-rack sprinklers and roof level sprinklers.
- ❖ Workers should be thoroughly trained in operation of fire alarms to prevent delays in evacuating the premises and calling the fire brigade in the event of a fire.
- ❖ Godown workers should give special trainings to operate firefighting equipment to deal with small fires.
- ❖ Seed godowns are often divided into compartments enclosed in fire-resisting construction, so such approach provides passive fire protection by inhibiting the spread of fire in the godown area.

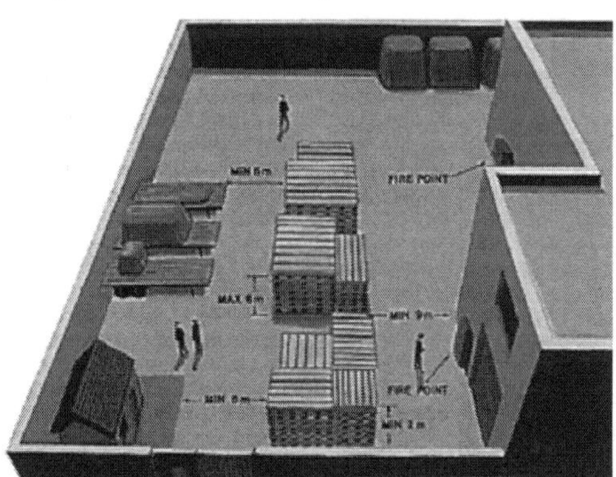

- Bulk storage of combustibles should be sited so as to, minimise the risk of fire occurring within the stack or spreading from adjacent areas.
- Preventing fires from occurring is as important as having properly working fire safety systems to deal with fire incident

Assessing Fire Risk

- The initial step in fire prevention is to assess the risks and record them in a risk register.
- Godown workers should review and assess the means by which a fire might start and spread, the potential consequences and available approaches to risk mitigation which includes day-to-day operations, risks with periodic building and maintenance work and those arising from installing new equipment, or adopting new or changing technologies.

Fire safety training

- Godown employees/workers should be given fire safety training from concerned person who is competent in subject and can understands effective training methods.
- Fire safety training should start with induction training on the day of appointment of new worker.
- Refresher training should be conducted at least once in a year to ensure that all workers/staff are familiar with fire precautions for workplace and are reminded of actions to take in case of emergency.
- More frequent training should be given where there is a high turnover of staff, or a high risk of fire.

Fire prevention

- To prevent fire, following principal actions need to be taken to monitor the behaviour of workers and prevent fires from occurring:
    - Smoking is one of the main fire risks and it should be prohibited in all godowns and workplaces and should only be allowed outside the godown area.
    - Enforce good housekeeping practices which covers implementing routines for regular removal and disposal of waste, establishing and maintaining out-of-hours inspection and security procedures
    - Doing routine checks, inspections and tests by monitoring the maintenance of heat generating equipment, chafing of cables, self-heating of cables due to electrical resistance and checks on fuel supplies and storage.
    - Issue and control work permits and associated procedures

- Instruct and supervise contractors and sub-contractors carrying out construction and maintenance operations in godown
- Avoid conditions leading to gas and dust explosion hazards
- Maintain integration with ventilation, communications systems

Inspection

- All fire protection equipment, including fire doors and partitions, should be inspected, tested regularly and be maintained in good working order.
- Fire doors should be kept closed at all times unless required to be opened for the movement of goods.

2

Stack Seed Bags/Bins/Containers

2.1 Receive seed bags/bins/containers in the receiving area and check that they are not open or damaged

2.1.1 Seed receiving area

- Seeds are not dispatched immediately to sellers or growers after being packed, but are kept for storage in godowns.
- Seeds before storage in godown, are first kept in seed receiving area in the godown.
- Seed receiving area is a place where seed bags/bins/containers after packaging and sealing are transported to godowns, where these bins/containers/bags are checked.
- Seed received at receiving area will be checked for any damage, leakage, seal opening, seed label etc and then will be send for storage or dispatch.

- Seed bags/container seal can be checked manually or by machines.

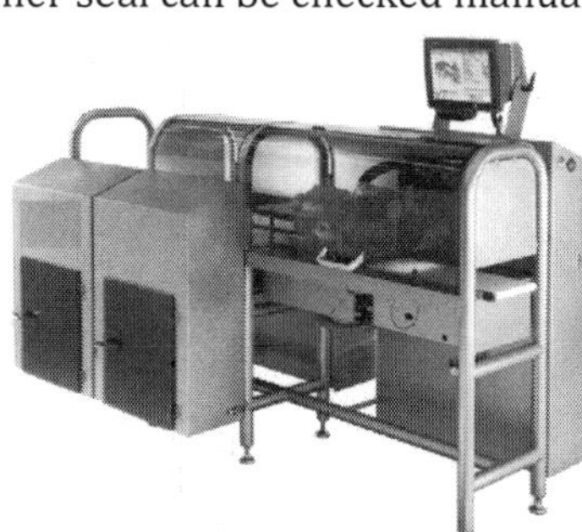

- If the seed bags/container/bin shows any of the above things, then godown manager will note for such missing/damage item, and such item will be send back.

2.1.2 Seed Storage/Dispatch

- Once the received seed bags/containers are checked, they are either stored or can be send for dispatch.
- In godown, seeds are stored on stacks or on pellets.
- For seed storage, godowns should have good storage capacity with sufficient height, proper light, ventilation, maintenance of internal temperature and allocation of separate slots for seed stacks etc.

- If the seeds bags are send for dispatch, then godown manager will enter the details in the register.
- The respective seeds bags quantities for dispatch are issue with gate pass or challan.
- Once the challan for respective seed bags quantities are issued, godown manager under his/her supervision will load the required seed bags/ container quantities in respective transport.

2.2 Stack bags/bins/containers on poles / wooden platforms leaving air space under them

2.2.1 Storing seed bags on wooden platform

- Depending on the size of storage, seed bags are normally formed into stack.
- Bags should be stacked on pallets or on an above ground structure to avoid the possibility of absorbing moisture from the floor.

- ❖ A one meter gap should be left between and around stacks and 1.5 m clearance between the top of the stack and the roof.
- ❖ Jute bags should not be stacked higher than 4 m while plastic bags should not be stacked higher than 3 m as plastic bags are more slippery and stacks will be less stable.
- ❖ Seed bags should not be stacked directly on floor, but should be stacked on wooden ballets.
- ❖ While keeping bags on wooden pallet, leave ample space around the stack, between stacks and the wall.
- ❖ The height of the stack should not be more than 6-8 bags.

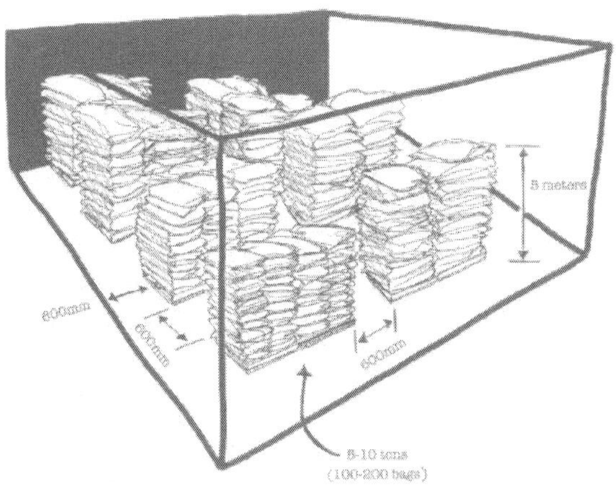

2.2.2 Precautions

Seeds bags should be laid out and stacked as per following:

- ❖ 1 meter distance between walls and stack
- ❖ 1.5 meters distance between roof and top of stack
- ❖ 1 meter distance between pillars and stack
- ❖ The air space between floor and pellet should be around 10mm.
- ❖ Access space should allow easy loading and unloading

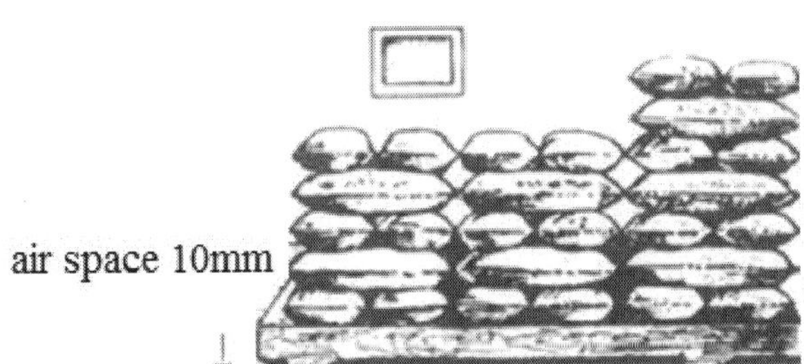

2.3 Keep bags in a way that air space is maintained between bags/ bins/containers and walls & roof of the store room / godown

2.3.1 Seed bag storage consideration

- Seed bags/bins should be stacked under the roof, in a shed or granary or under water proof tarpaulins.
- For storing seed bags, atleast 600mm gap should be left between and around stacks and 1.5 meters clearance between the top of the stack and the roof for proper air circulation and pest management.

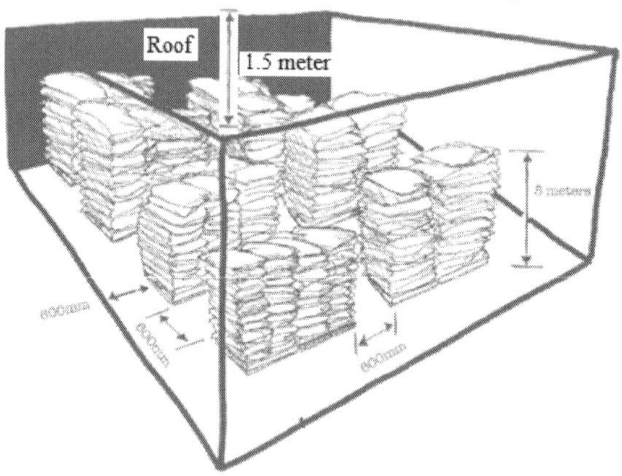

- Seed bags/containers should be stacked on pallets or on an above ground structure to avoid the possibility of absorbing moisture from the floor.
- Seed bags should be stacked so that fumigation can be undertaken easily as dimensions of stacks should be set to facilitate sealing with single fumigation sheet.

2.4 Keeping bags in upright position or as mentioned in the SOP

- The received seed bags should be kept on wooden pallets and should be arranged in orderly pattern to facilitate counting and inspection.
- Seed bags should be kept at upright position for proper ventilation.
- While storing seeds in upright position, it is important to leave adequate space for air to move quickly.

2.5 Stack different seeds lots separately

- The processed seeds are handled in bags which are stored in the stack.
- Seed lot separation is important for identification purposes and in preventing contamination.
- Bags of seeds of each genus, species and cultivar are stored in separate stacks.

- For proper ventilation, stacks of seed bags should be well spaced, between bags and stacks.
- Bags of seeds must be stacked properly to prevent slippage.
- Stack numbering layout should be displayed in each godown clearly by labeling the location and number of each stack.
- Seed processing worker should see that seed bags/container for different cultivars should be labeled properly.
- Unloading of seed should be done in individual stacks, with at least 1m gap with next immediate stack or wall for proper movement, cleaning/house-keeping and for proper sealing of stack for any future fumigation.
- Stack the seed with tarpaulin sheet on floor, with sheet extending at least half-a-meter beyond the stack.
- Each material/lot should be stacked separately, ensuring no mixups.

2.6 Ensure that no other feed; fertilizer; salt; medicines or chemicals are stored alongside seed bags / bins / containers in the store room / godown

- Chemicals, feed, fuels, fertilisers should not be stored along with seed bags/ container in store room.
- Seed or godown worker should see that fertilizers should not be stored with chemicals in dry conditions.
- As recommended, pesticides should be separated with at least one pallet width or three feet from food, feed and seeds.
- Seed storage areas should not contain pesticides or other chemicals;

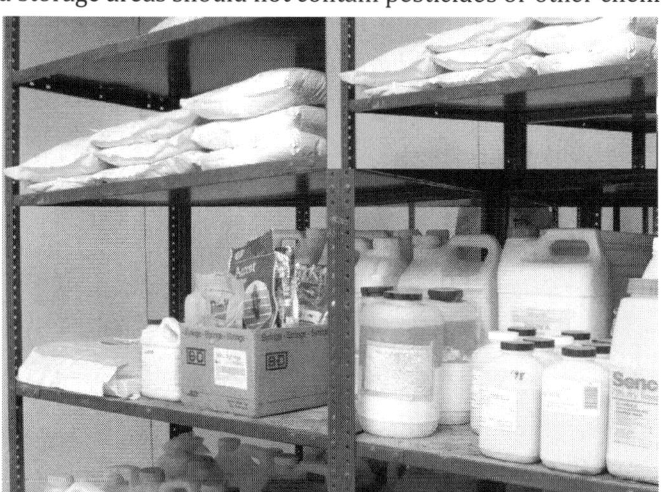

2.7. Keep complete and up-to-date records of the seed inventory as per the SOP

- Inventory control system in godowns includes all aspects of managing seed bags inventories right from purchasing, shipping, receiving, tracking, warehousing, storage, turnover till reordering.

- Inventory manager/seed worker shall keep complete details about all seed bags/containers/bins as they are unloaded or loaded since such information needs to be attached with delivery ticket in case of any notations on damage or shortage.
- The inventory manager/ seed worker keeps track of seed that moves in and out of particular warehouse location.
- Godown manager should maintain accurate inventory and should keep accurate records of incoming and outgoing seed stock on bin card.
- Apart from entering details in bin card, inventory manager/ seed worker is also responsible for noting seed bag price labels or to record barcode information in computer.

Inventory manager/seed worker should keep following updated information about seed stock on his/her log book/computer:

- Accession number (number).
- Introduction ID number (number).
- Collection name, pedigree (name and number).
- Associate number (other genebank's accession numbers).
- Collection site (country of origin, state name, longitude, latitude, altitude).
- Donor (donor institution).
- Race class (maize race, race purity, common name).
- Grain type (texture, colour).
- Origin of the stored seeds (location, year, field plot, sample).

- Photo [sample photo of ear and/or kernels (grain type: colours and textures)].
- Reference sample (location in the bank, seed origin).
- Seed exchange policy (MLS, specific).
- Seed health (cleared for outgoing, not cleared).
- Seed shipment (collaborator, institution, country).
- Safety duplicate storage information (collaborating genebanks).
- Seed monitoring (seed amounts and seed viability).
- Accession availability status (available, not available).
- Storage address (row-column-level-sample).
- Amount of seed stored (seed number or seed weight).

3

Achieve Productivity and Quality Standards

3.1. Ensure storage of different seed lots separately

- Seed bags should be stenciled with lot number and should carry a label showing much information.
- The lot number of each seed lot corresponds to lot number of container labels.
- Each seed bags/container of incoming seeds should be labelled, so that identity can always be maintained and easily determined.
- Lots of same seed variety and similar quality will be stored in specified seed storage area in the godown.

Correct position of Seed Lots

In-correct position of Seed Lots

- For the proper maintenance of seeds in godown, different seed lot should be kept separately

3.2. Ensure that stack bags of one lot are not on top of a different lot

- Seed processing worker needs to make sure that while stacking of seed bags/containers, lots of same seed lot, variety and quality should be kept on similar seed lot.

- ❖ Since each seed bags/container of incoming seeds are labeled, so seed worker as per the label seed lot details should keep the respective seed bags.
- ❖ Seed processing worker while staking seed lot should also verify seed bags from lot number.

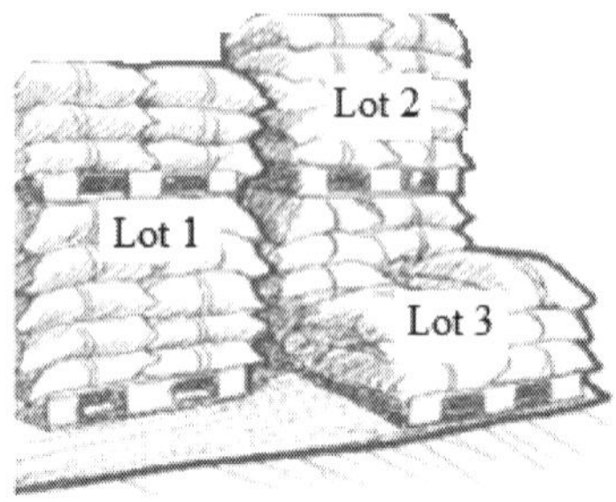

Correct position of Seed Lots

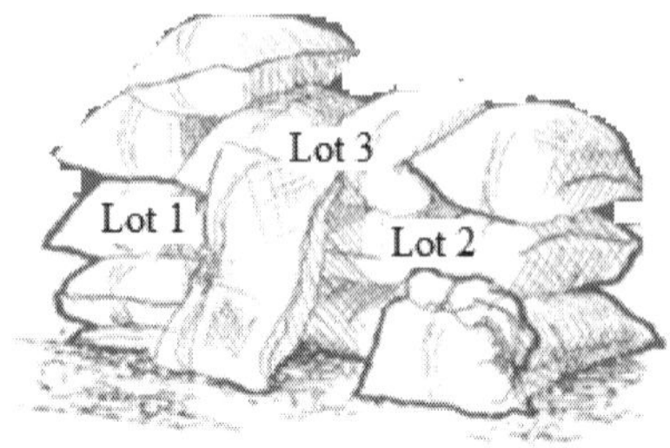

In-correct position of Seed Lots

3.3 Ensure stacking of bags to any efficient storage height without causing weight or pressure damage to seed at the bottom

- ❖ Proper stacking is important in seed storage as it ensures efficient use of storage capacity and prevents damage to seeds.
- ❖ Properly dried seed bags needs to be stacked to sufficient storage height without weight or pressure damage to seeds at the bottom as excess height lead to mechanical damage of seeds.
- ❖ Seeds inside seed bags also gets damaged from mechanical injuries due to falling or dropping of seed bags.
- ❖ The height of stack is determined based on goods, size, weight, shape, strength of packing to stand the height of stack.
- ❖ Also, the height of stacks is conditioned by type of bags (plant fibre or plastic), by fragility of the products and by very dimensions of storehouse.
- ❖ Jute bags should not be stacked higher than 4 m while plastic bags should not be stacked higher than 3 m as it will help in seed damage due to weight.
- ❖ While storing bags on pallets, do not exceed the maximum recommended stack size in terms of number of bags and height of stack.

For keeping the bottom seed bags safe, seed processing worker needs to check that:

- ❖ stacking of seed bags should be done in pattern which prevents falling or sliding
- ❖ staking of bags should not be more than six layers on a pallet

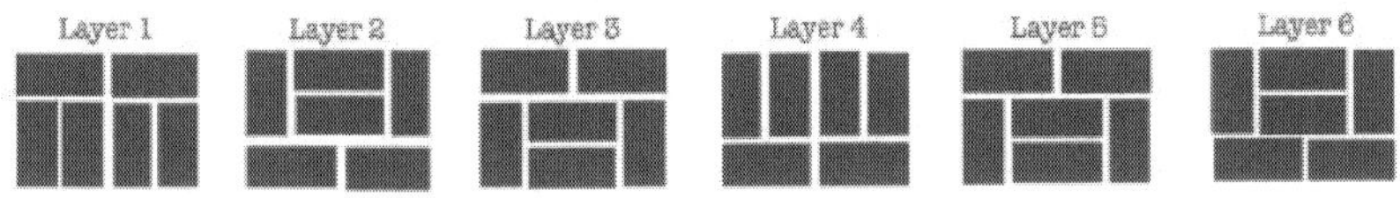

- staking of jute bags should not be more than five meters
- staking of plastic bags should not be more than three meters, as such bags are slippery
- overlapping and interlocking of seeds bags should be there in every layer
- height of the stacks should not be greater than its width.
- for stability of stack, each bag should be stacked so that it always covered by two other bags.

3.4. Ensure that different area of store room / godown are properly marked for easy traceability of seed bags/bins/containers

3.4.1 Traceability

- Traceability is an ability to track and trace commodities/seeds across multiple trading partners with real-time visibility and quickly respond to market needs for success and efficient supply chain management.
- As seeds are packed in relatively small boxes to minimize damage, implementation of traceability has also been relatively low cost.
- Seed traceability has two systems of information, where first involves with physical labels on seed bags/boxes and on pallets as it is important to identify the product in bags/boxes while other involves paper or electronic trail to trace back between different links in marketing chain as each link uses different traceability system.

3.4.2 Importance

Traceability is important in supply chain as it will help in:

- knowing current location of the seeds
- rapid withdraw harmful or unsafe seeds from market
- better information to consumers about seeds, its origin and method of production
- improving stock management and return flows
- preventing counterfeit of seeds
- enhancing consumer confidence by enhanced seed safety
- better expiry date and waste management
- improved Product Life Cycle management
- ensuring right product for right use

3.4.3 Traceability of seed bags in godowns

- Seed bags/container before sealing are marked with 2D Bar-Code which contain different levels of information which will be easy for seed packets/bags/canes traceability.
- The seed bags before stacking are stored as per stack label so that proper seed lot identification can be done for future traceability.
- Each seed lot/bags should be labeled accurately and should be kept in required stack layer so that godown manager can easily trace the required seed lot and can correctly maintain the details in computer and in registers.

- Each stack shall be identified by drawing a 5cm width stack line in yellow or white colour on all four sides of the floor of godown as per stack plan.
- The stack shall be given stack number which can be painted on the floor / wall / pillar in front of each stack for proper traceability of seed bags/containers.

3.5 Ensure that bags are not dropped-off during handling

3.5.1 Seed bag storage

- After seeds are dried, cleaned and treated with suitable chemicals they are packaged in bags/containers.

- Seed bags are stored in piles, single layers, sacks or in open racks at certain heights.
- If proper stacking of seed bags are not done, they can be damaged due to rough handling.

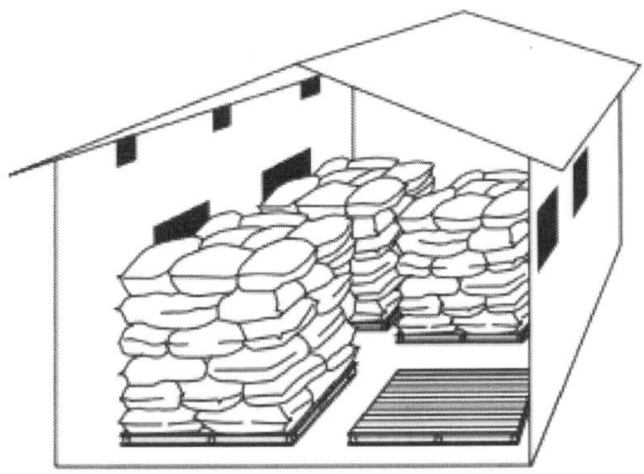

- Seeds bags should not be thrown or dropped as will lead to cracking of seed coats and embryos which results in seed that won't develop properly.
- Seed bags must be stacked in such a way to prevent movement during transportation and should be so secured that, they will not fall off when corners or inclines are being negotiated.

3.5.2 Precautions

Godown manager or seed processing worker should make sure that following points should be consider while handling seeds bags/containers:

- While lifting synthetic bags or sacks, the godown worker should take certain precaution as such type of bag material has a tendency to slip when new.
- Godown workers while storing/keeping the seed bags should keep the bags in flat position one above another in interlocking manner, so that bags will not fall.
- Seeds bag material of different sizes and shapes should not be stacked on top of each other unless proper precautions are taken to prevent movement by settling or vibration.
- Seed storage areas should be kept clean and free from all refuse and incorrectly stored materials as it will give trouble to godown workers, who carry loads on them.
- Sufficient and suitable lighting must be provided in every part of undertaking where workers are employed or pass.
- While lifting seed bags, godown employees must wear gloves and boots to avoid any droppage or slippage.

3.6 Ensure that the storage place is spotlessly clean all the time

- Good hygiene in seed godown is important in maintaining grain and seed quality.
- For maintaining good hygiene in storage:
  - Keeping seed storage areas clean which includes sweeping the floor, removing cobwebs and dust, and collecting and removing grain spills.
  - Seed storage area should be clean once emptied by spraying walls, crevices and wooden pallets using insecticide before using them again.
  - Keep rats and mice away.
  - sealed storage room or seed stock with tarpaulin and treat then with fumigants.

3.7 Ensure proper documentation as per the company's SOP

In order to have complete documentation about seeds and its packages, seed processing worker should keep complete information about various seed documents which includes:

Seed source documents

Such documents contain information on:

- Name of seed species (botanical and local names)
- Geographical location of seed source (elevation, temperatures, rainfall, soil types)
- Geographical location of original seed source in case of seed orchard or seed production unit
- Ecology of seed source
- Ecology of original seed source in case of seed orchard or seed production unit
- Number of trees in seed source
- Age of seed source
- Size of seed source

Seed collection and handling documents

Such documents contain information on:

- Names of the species (botanical and local names)
- Seed source or provenance (from where the seed was collected)
- Date of seed collection
- Number of parent trees from which the seed was collected

- Average distance between seed trees
- Weight of fruit or seed collected
- Number of seed containers filled with the seed collected
- Name of collectors

Seed storage documents

Such documents contain information on;

- Seedlot number (a unique identification code to identify the seed collected on a specific year from a specific seed source)
- Names of the species (botanical and local names)
- Weight stored
- Running balance

Seed Dispatch

For information on seed dispatch, such documents contain information on;

- Names of the species (botanical and local names)
- Weight required
- Weight issued
- Seedlot number (a unique identification code to identify the seed collected on a specific date from a specific seed source)
- Date of dispatch
- Seed test information (purity and germination)
- Seed advice note
- Dispatch note.
- No of parent trees

Seed Labelling

For information on seed dispatch, such documents contain information on;

- Botanical and local name of species
- Date of collection
- Seed source/provenance
- No of parent trees
- Seed quantity (Kg)
- Purity percentage
- Viability test result or germination percentage:
- Method of pre-sowing treatment

Unit 5: Maintain Safe Work Environment

1

Maintain Standard Safety Procedures at the Workplace

1.1 Comply with general safety procedures of the company

1.1.1 Safety practice at workplace

Safe practices and safety conscious attitude by staff is the best protection against workplace accidents. To help in assessing safety status of seeds business, seed workers should answer the following questions, highlighting what's missing and determining where he/she needs an improvement. By completing such questions, seed workers can able to monitor his/her workplace health and safety performance which ensures his/her regular improvement.

Proper safety is important to any operation, as day-to-day activities of production/sales can take in focus off things like safety. To fight with combat that, seed workers should make sure that he/she should regularly talk about safety practices and procedures with owners, managers and workers.

Following are the points which should be managed to handle safety procedure for employees/customers at workplace:

1. Dry and clean slip proof floors/pathways:

- Slips and falls are one of the common ways of enduring injuries, so to prevent injuries, the floors/pathways should be cleaned on regular basis.
- Slip can also occur due to wet contaminants such as water, oil or grease, or dry such as dust, metal, plastic bags or off-cuts.
- Preventing floor contaminants is one of the best things you can do to prevent slips.
- More serious slips or trips together with the resulting falls may result in:
  - sprains or strains
  - broken bones when trying to break the fall
  - a back injury due to the sudden and forceful impact during a fall

- burns if it occurs near hot surfaces or if the person is handling hot fluids
- cuts if it occurs near sharp objects.

2. Unorganized and messy environment

- The messy environment is the main cause of all problems.
- In this, there are more chances of threats going unnoticeable and the situation may simply turn the place into accident-prone area.
- An occupational hazard is a result of ignorance on part of workers or carelessness on part of workers.
- If workplace is messy and unorganized, it may be a door to many uninvited hazardous accidents, so a gardener should properly arrange and clean the working area from tools, chemicals, fertilizers etc.
- Proper piling of pots, fertilizer bags, and chemical boxes should be done in the store area, otherwise locating the items from improper piling will lead to stack falling upon workers and can make him/her injured.
- An unorganized store may also hamper the productivity rate of the worker as for locating particular item from the store without proper stocking will lead to time wastage of worker.

3. Use of disinfectants in workplace

- Disinfectants are very important to be used in garden, parks, public places as there are more risk of germs spreading around due to multiple uses of each and every item.
- In workplaces, there is frequent use of items from one worker to another, whether it is flu season or not, as germs are inevitable in nature.
- The disinfectants comparatively lessen the chance of germ produced illness or diseases.
- In order to cut the cost some workers opt for cheap disinfectant, which on longer run can affect the health and in turn, will lead to health issues and absentees.

4. Germ controls to fight flu/diseases:

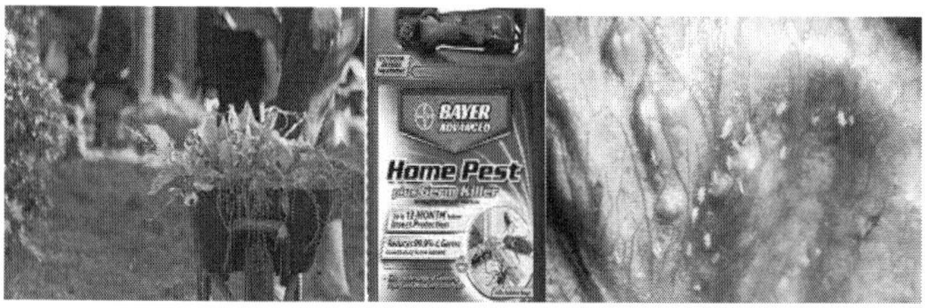

- Germs can easily disperse across the entire workplace if there is any compromise in the cleanliness.
- The sitting place or lunchroom can be called as germ house for workers, so it is important to clean and sanitize such areas frequently.
- Apart from flu, there is high chance of catching fatal disease or infection by shared germs in the workplace.
- There are many infectious germs that may spread due to unhygienic environment and lead to various disease such as fungal infection which can lead to irritation and skin infections.

5. Green and clean equipments/products

- Eco-friendly products are best in nature and are not harmful to human health.
- Such products are friendly to nature.
- Not all cleaning products can be compared to one another as all such products have different standards.
- Some products are cheaper and harmful whereas some are costlier and trustworthy.
- It is very important to choose the correct chemical.
- While choosing chemicals, it is equally important to store these chemicals in correct places.
- There are more chances of such chemical to cause skin related allergies, which on longer run can affect the worker's attention span and confidence.

6. Air filtering and humidity

- Airborne diseases do not stay back in race of harming and spreading more and more diseases.
- One may not be able see them, but they can be equal threat in the way of the organization at a whole.
- A better ventilation system works as backbone against airborne diseases.

- Germs that can pass through respiratory infections can be tackled by ventilating system.
- It will not only enhance the productivity of workers but also maintain their good health.
- Thus, regular cleaning and timely changing of the required filters is very much advisable. Germs restrict the potential microbial growth or odor related problems in work environment.
- Humidity affect the germ level in workplace, so maintaining proper humidity level in store room or sitting area is advisable.

7. Proper lightening

- Dirty light fixtures are indeed one reason to inefficient light supply.
- It can be a great hassle in way of workers.
- It is proven that adequate light is required in order to do daily task.
- No proper lightening can be an unsafe combination with workplace.
- Pathways, store rooms and equipment areas should be well lighted to prevent accidents and contribute to healthy and safe environment.

8. Perfect waste disposal plan

- Piled up trash will result in dirt all around the workplaces.
- Piled trash is a place where pesticides can reside and multiply.
- Pesticides can be a great threat and hazard to working environment.
- Placing dust bins or waste bins throughout the workplaces is a way to gather waste in a single place.

- Proper collection of waste from dustbins should be followed so that the waste will not spread anywhere in working area.
- Seed worker should implement reducing and recycling methods in workplace.

9. Image is everything

- Workplace should be clean as apart from workers, there are customers, suppliers or sellers who used to roam in workplace.
- If the environment of workplace is messed up, unorganized and dirty, then there will be less possibility of good customers to visit and buy the product.
- Clean workplace will give better impression to the customers and will lead to good product sale.

1.2. Follow standard safety procedures while handling equipment, hazardous material or tool

1.2.1 Introduction

Workplace emergency is an unforeseen situation that threatens your employees, customers, or the public; disrupts or shuts down your operations; or causes physical or environmental damage. An emergency response plan covers designated actions employers and employees must take to ensure employee safety from fire and other emergencies.

A worker's attitude is the main cause with regards to machinery safety. Accidents at site do happens and often do occur due to human error, so safety hazards should be identified by pesticide and fertilizer applicator in all areas of machinery use. Seed workers must be on constant alert to possible hazardous situations in machinery operation, seed dispatching, seed packaging, seed bags loading and handling. Accidents can be reduced if workers would stop and think about risks involved in handling equipments, hazardous materials and using tools.

1.2.2 Risk at workplace

Risk is a chance that someone could be harmed by hazards like working from ladders or working around electricity. A risk assessment is an important step in protecting people, business and complying with applicable laws.

Farmers/seed workers are particularly exposed to ergonomic hazards such as:

- ❖ tiring or painful positions
- ❖ carrying or moving heavy loads
- ❖ standing or walking and repetitive hand or arm movements
- ❖ work-related musculoskeletal problems

Seed workers live with risk and make decisions every day that affect their operations. Many factors that affect the decisions which seed workers make could not be predicted accurately such as:

- ❖ change in weather conditions
- ❖ drop in prices at time of harvest
- ❖ non availability of hired labour during working times
- ❖ breakage of machinery and equipment when required
- ❖ dying of seeds
- ❖ change in government policies

The above changes are examples of risks which seed workers face in managing their seed business which will finally affect seed business profitability.

1.2.3 Injuries due to machinery/vehicle

Seed/ godown equipments and vehicles daily check are simple and effective way to spot potentially dangerous issues or defects before they are used. Seed/godown workers should need to have a system of daily routine checks for machineries and equipments that are used in processing, loading, harvesting and cleaning areas to ensure that are in good working order, safe and fit for purpose before start of working day or work shift.

Seed workers uses froliks, elevators, weighing machines and scales, pelleting machines, cleaning machines, separators, cylinders equipment every day. Such machines are powerful which helps seed workers to work more quickly and efficiently, but such machinery brings new dangers.

The main causes of risk/accidents due to equipments while working at site include:

- ❖ transport (running over or vehicle over turnings);
- ❖ falling from height (from fragile roofs, stacks etc);
- ❖ struck by moving or falling objects (bales, trees etc);
- ❖ trapped by something collapsing or overturning;
- ❖ livestock related accidents and fatalities;
- ❖ chemical poisoning

Poorly used or faulty machinery is main cause of injury on seed areas. Machinery accidents can be prevented by keeping the machine in good repair, fitting and ensuring all safety equipment like guards, safe access platforms and ROPS to be operational with machine all the times.

Seed workers before using the machines should ask themselves:

- ❖ Is the machine they intend to use is suitable for the job?
- ❖ Are all safety devices like guards are in place and are working correctly?
- ❖ Are machine operator properly trained to do this job/use this machine?
- ❖ Has the instruction manual for the machine been provided, read and understood?
- ❖ Is the right personal protective equipment is available and worn?
- ❖ Has a risk assessment been carried out?
- ❖ Has the work been properly planned and communicated to those who may be at risk?
- ❖ If two or more people are involved, has everyone understood what needs to be done and has a system of communication been agreed on?
- ❖ Are all operators/workers trained to do this job?
- ❖ Is suitable clothing and footwear available and worn (including any PPE)?
- ❖ Have operators been provided with relevant information?

1.2.4 Managing equipment injuries

Industry standards give reasonable degree of personal safety for operators and others during normal operation and servicing of equipment. Seed working and handling equipments should be inspected to find whether the equipment can be operated, adjusted and maintained safely with any deterioration detected and remedied before it results in health and safety risk. Not all working equipment needs formal inspection to ensure safety, but in many cases, quick visual check should be required by seed workers before using such machinery.

Inspection is necessary for any equipment where significant risks to health and safety may arise from incorrect installation, reinstallation, deterioration or any other circumstances. The need for inspection and inspection frequencies should be determined through risk assessment, so before using, seed workers should inspect/ check the following equipments:

- ❖ elevators,
- ❖ weighing machines and scales,
- ❖ pelleting machines,
- ❖ seed cleaning machines,

- separators and cylinders,
- chemical cans

Inspection of equipment should concentrate on those safety-related parts/ procedures which are necessary for safe operation of work equipment which needs testing or dismantling. However, not all safety-critical features on a particular item of work equipment may require inspection at the same intervals. Such inspection can vary in its extent, as the following demonstrate:

- quick checks before use
- weekly checks
- more extensive examinations, undertaken every few months or longer

1.2.5 Risk due to Hazardous materials

Pesticides are chemicals that deter and eliminate certain pest populations including insects, rodents, fungi, weeds, and other animals. The increased quantity and frequency of pesticide applications have posed major challenge to targeted pests causing them to either disperse to new environment and/or adapt to the novel conditions. The adaptation of pest to new environment could be attributed to the several mechanisms such as gene mutation, change in population growth rates, and increase in number of generations etc.

Pesticides have also been reported to influence mineralization of soil organic matter, which is a key soil property that determines the soil quality and productivity. Pesticides that reach the soil may also disturb local metabolism or can alter the soil enzymatic activity.

1.2.6 Safety procedure for handling hazardous materials

There are certain basic rules which all workers who handle hazardous materials should know and follow which includes:

- Follow all established procedures and perform job duties as you've been trained.
- Be cautious and plan ahead and think about what could go wrong and pay close attention to what you're doing while you work.
- Always use required protective equipment and inspect it carefully before each use to make sure it's safe to use. Replace worn out or damage PPE; it won't provide adequate protection.
- Make sure all containers/bags are properly labeled and that the material is contained in an appropriate container. Don't use any material not contained or labeled properly. Report any damaged containers or illegible labels to supervisor.

- Read labels and material safety data sheet before using any material to make sure you understand hazards and precautions.
- Use all materials solely for their intended purpose. Don't use solvents to clean your hands, or gasoline to wipe down equipment.
- Never eat or drink while handling any materials, and if your hands are contaminated, don't use cosmetics or handle contact lenses.
- Read the labels to identify properties and hazards of products and materials.
- Store all materials properly, separate incompatibles, and store in ventilated, dry, cool areas.
- Keep you and your work area clean and after handling any material, wash hands with soap and water. Clean work surfaces at least once a shift so that contamination risks are minimized.
- Learn about emergency procedures and equipment by understanding evacuation procedures, emergency reporting procedures, and procedures for dealing with fires and spills.

1.2.7 Managing Risk

Reducing risk is often used in combination with one another as no single strategy can cover all of risk likely to be encountered. Seed workers need to consider the risks simultaneously and to develop an integrated approach for better management. They need to recognize the advantages and disadvantages of each risk management option both individually and in combination. Each workers should select an appropriate strategy based on their goals, attitudes towards risk and their personal and financial situations.

Seed workers should choose and combine strategies based on their goals, attitudes towards risk and their personal and financial situations. Every seed worker family required to find their own ways of coping with risk as workers responses to risk are as diverse as risks that affect their working area.

Seed workers reduce risk by learning about and applying new technologies and practices designed to address specific risks common to their area of production such use of:

- drought resistant seed
- bird resistant seed
- disease and pest-resistant seed species
- disease resistant livestock species
- simplified loading of seed bags
- correct seed storage
- proper seed handling to improve marketability

1.3. Check parts of workplace and take preventive actions like spraying and other steps to protect from leakages, water logging, pests, fire, pollution, etc.

1.3.1 Introduction

- Seed godowns contains storage pests like rodents, beetles and moths which damages seeds.
- Seeds are packed in jute, plastic and polythene bags and are sometimes kept in plastic containers which are suspect to fire in seed godown.
- Seed godowns are normally permanent structures which are often made of bricks and asbestos sheets which can lead to water leakage.

1.3.2 Checking seed damage by:

Pests

- Seed storage pests like insects, pathogens, rodents and birds are the main cause of seed damage in godowns, so seed processing worker should take care of:
- Timely detection of storage pests as it plays major role in seed control measures.
- Seed storage pest cause losses by combination of feeding, spoiling and contamination of both paddy and milled grain.

For preventing seed pests, effective pest control should be used which involves:

- Harvesting, drying and storage of clean dry grain
- Disinfecting the storage system and
- Controlling or preventing pest infestation during the storage period.

Rats and mice

Rats and mice cause losses in a number of ways:

- Feeding on stored produce
  - Rats consume about 25 gm of seed in a day and mice eats around 3-4 gm in a day.
  - Besides eating stored seeds, rodents contaminate a lot of stored produce with urine, feces, hair and pathogenic agents.
  - As it is extremely difficult to remove contamination, infested batches often have to be declared unfit for human consumption.
- Damage to material and equipment
  - Produce leaking out of damaged bags or storage containers
  - Bag stacks collapsing due to damage to lower layers

- Short circuits leading to sparks or fire from cables being chewed
- Silos and warehouses may subside or even collapse as a result of being undermined
- Drainage canals around a store may be damaged.

Preventive measures against Rodents

- The most essential factors for occurrence of rodents are:
  - Sufficient supplies of food
  - Protected places in which to build burrows and nests
  - Hiding places
  - Access to produce
- Good store management and preventive measures taken as part of an integrated control program can help to deal with such factors which includes:
  - Keeping godowns absolutely clean by removing any spilt grain immediately as it will attracts the rodents.
  - Store bags in tidy stacks set up on pallets, and ensures that there is a space of 1m all round the stack.

1.3.3 Checking cause of fire

Understanding the common causes of godown fires puts seed worker in good position to do best possible work, what he/she can do for fire safety and preventing fire before they have opportunity to start. There are certain common causes which predict the cause of fire in godown such as:

- Fires which sets intentionally: It is an unfortunate case that by far one of most common causes of warehouse fires are those that are deliberately started for taking insurance.
- Problems with electricity and lighting: It is a common causes of warehouse fires which are less damaging to properties than intentionally started fires, but they are still a major problem for those in charge of warehouse fire safety. This is something that those in charge of warehouse safety need to take very seriously.
- Heating equipment: In cold, godown workers uses certain equipments which can cause serious fire risk as more that 8% of warehouse fires are caused by heating equipment.
- Exposure: It refers to materials that are combustible but that are not currently on fire, so warehouses that contain large amounts of materials that could catch fire if they were exposed to a flame or heat source.

- Smoking materials: It is important as materials involved in smoking causes large number of warehouse fires which includes cigarettes, cigars and lighters.

Preventing Fire

There are many things that seed business can be to minimise their risk of fires occurring which can be:

- Fire risk assessments: It is a legal requirement to have fire risk assessment carried out under Regulatory Reform, however not all assessments are created equal. It is advisable to seed worker/manager to have professionals manage such process to ensure that it is carried out correctly.
- Smoke detectors: Seed worker should ensure that he/she have smoke detectors installed across the warehouse paying special attention to any areas that may be vulnerable to fire breaking out.
- Fire extinguishers: seed worker should have appropriate fire extinguishers in place across the warehouse, however it is important to note that fire extinguishers should only be used appropriately.

1.3.4 Causes of water leakage

- Seed worker should be extra vigilant when it comes to protecting seed bags/containers against expensive water damage.
- Incoming water is more than nuisance as it can be equivalent to huge repairs bill or adding damage slowly over time.

The water problem in godowns can be due to:

Mold and mildew

- Unseen water or high humidity can result in mold and mildew.
- Neither is as innocuous, but in fact, both pose safety hazards, and removal can be incredibly expensive.

Cracks

- A godown is prone to damage that can be made worse by presence of water through rains.
- Tiny cracks in walls, ceilings and floors allow moisture to enter in godowns.
- As temperature changes, moisture can be wrecking ball for structure as it expands and contracts.

Large-scale damages

- A burst pipe or extreme weather can wreak havoc on godown in short amount of time.
- It can do structural damage and destroy seed bags/containers.

Precautions

- The godown should have roof which should overhang the gables by 0.7 to 1.0 metres, and eaves by at least 1 metre to ensure that rainwater is shed well clear of walls.
- The overhang also helps to keep godown walls cool and protects ventilation openings from rain.

1.4. Ensure no accidents and damages at workplace, reporting of any breach of company safety procedure

1.4.1 Introduction

- A workplace accident can be described as an occurrence during work / in workplace that leads to injury.
- A workplace accident is made when workers can prove that the accident in question was a direct result of negligence at the hands of processing worker/workers.
- Accidents at work can happen in multitude of scenarios, to anyone and at any time, so it is important to always be vigilant of hazards and follow safety procedures correctly to avoid any accidents taking place.

1.4.2 Types of Accidents

1. Slip, Trip and fall

Slip, trips and falls are common causes of personal injuries. At workplace, workers are likely to encounter a slippery, uneven or hazardous surface at some stage. Such accidents are completely avoidable where correct safety measures are in place and employer may be found liable if it is shown that they were negligent in ensuring a safe working environment for their employees.

2 Manual handling

Manual handling is another common workplace accident where back, neck and shoulder injuries are common for those who lift objects as part of their daily tasks at work.

Manual handling covers a wide range of activities including:

- lifting,
- pushing,
- pulling,
- holding,
- restraining,
- throwing

- carrying

Manual handling also includes repetitive tasks such as:

- packing,
- filling,
- assembling,
- cleaning
- sorting

It is seen that not all manual handling tasks are hazardous, but as most jobs involve some form of manual handling, most workers/applicators are at certain risk of manual handling injuries. Good posture and lifting techniques will help in reducing risks, but research shows that making changes to workplace design is most effective way to prevent manual handling injuries.

3. Hit by Falling Objects

In all environments, whether it warehouse or an office, injury from falling objects can happen. Falling objects can cause injuries such as cuts and head and brain injuries in severe cases. Common causes of falling object injuries are:

- Files, books, stationery and other office tools and equipment falling from a shelf
- Wall and ceiling fixtures become detached and falling
- Tools and other construction materials and debris falling from a height on a building site
- Unsecured loads lifted to a height and falling
- Collapsing shelves that haven't been properly maintained

4. Injuries by inadequate Personal Protective Equipment (PPE)

Accidents in workplace tend to happen when processing worker fail to provide appropriate protective equipment for seed workers. Processing worker have a duty of care to their seed workers to provide adequate PPE to those who need them. In cases where processing worker fails to provide adequate equipment, it will lead to breach in their duty of care and when this happens, injuries occur. Some common examples of inadequate equipment are:

- Eye goggles that slip off
- Safety boots that don't fit or are not suitable for the work environment
- Hard hats that don't fit correctly
- Gloves that don't fit or are worn or have holes in them
- Protective masks that are worn or have holes in them

5. Falling from Height

Falling from height is common in godowns as seed workers needs to keep the seed bags on racks using ladder or frolic. Falling from a height usually means that a person was either

- working on a ladder,
- a roof,
- using a step ladder to reach a high place,
- working on a fragile surface,
- falling from a high vehicle.

Common examples of causes of falls from heights in work are:

- Falling from scaffolding on a building site
- Using a step ladder when it is not appropriate to do so
- Overreaching and losing balance when on a ladder
- Not ensuring the ladder is properly footed

6. Walking into Objects

Walking into objects is common occurrence as workers become distracted and simply walk into an object of which he/she didn't expect to be there, either because they were distracted by another task or object which was not supposed to be there and obstructed their walkway.

7. Burn Injuries

Burn injuries in workplace can be caused by fire and explosions due to faulty wiring and also from contact with corrosive materials, electrical burns and chemical burns. Common causes of burn injuries in workplace are:

- Electrical burns from contact with electrical currents caused by defective tools or faulty electrical equipment
- Contact with overheated machinery or tools
- Chemical burns from contact with corrosive materials, acids or oils
- Contact with direct heat sources such as welding.
- Fire burns from a fire started within the building or by faulty wiring.

8. Repetitive Strain Injuries

Reparative strain injury is a disorder which causes discomfort and/or pain in muscles, tendon and nerves and is caused by repetitive motion or overuse. It is generally a condition that affects the upper body (hands, wrists, fingers, elbows, neck, shoulders). It can happen to manual workers such as those is a warehouse or on an assembly line but can also happen to those who work in front of a computer continuously. Common causes are:

- Overuse
- Repetitive tasks
- Inadequate break times during the day
- Poor posture
- Badly designed office chairs and keyboards
- Repeated use of vibrating tools
- Heavy lifting

9. Vehicle Crashes and Collisions

These type of workplace accidents may involve cars, lorries or even other vehicles such as forklift trucks. These accidents include:

- being struck by a moving vehicle,
- being struck by falling objects from a vehicle,
- falling from a vehicle or getting stuck or crushed by overturned vehicles.
- There are two distinct kinds of vehicle-related accidents.

10. Dangerous Machinery Accidents

Most warehouses use dangerous machinery daily. Common causes of accidents involving dangerous machinery are:

- Eye injuries caused by dislodged parts or debris from machinery
- Machine entanglement and crush injuries
- Cuts and lacerations from coming in contact with sharp tools or parts of machinery
- Lack of training of employees
- Faulty or inadequate personal protective equipment
- Faulty machinery or poorly maintained machinery

1.4.3 Safe Practice

While working in seed godowns, seed processing worker should know how they have to work in order to handle task which are hazardous and which involves risk of injury by following the safe working practice and procedures such as:

- Avoiding rushing of work
- Avoid using brute force
- Encourage seed worker to lift the material/equipment by sharing with each other

- Asking seed workers to change their posture regularly and take move and stretch among their tasks.
- Ask seed workers to stay themselves fit.
- Encourage seed workers for early reporting of any serious issues.
- Ask seed workers to stay active and keep moving which helps in building strength gradually.
- Avoiding heavy duties.
- Take medication immediately.
- Ask seed workers not to lift the objects that are too big or heavy.
- Use wheeled trolleys, wheelbarrows, ramps and other mechanical aids.
- Work to be planned in advance so that workers don't have to work for extra hours.
- Ask seed workers to wear personal protective equipment before doing any hazardous work.

1.5. Keep workplace organized, swept, clean and hazard free

1.5.1 Introduction

- Cleaning is a process of getting a clean, free, and energetic environment where there is no risk of any hazard.
- Clean surrounding makes it easier to organize the workplace so that it is easy to access all sort of material which are required.
- Cleanliness is important quality of life and serves as a habit that is often considered just next to godliness.
- Cleanliness is nothing that should be forced on a person but should be properly practiced and be made a lifestyle.

- Cleanliness is required for health and wellness of an individuals and society.
- Health and cleanliness relates to each other and influences each other. In order to be in good health, one should practice hygiene.
- Hygiene is the practice of maintaining good health and preventing diseases through cleanliness.

1.5.2 Keeping workplace hazards free

- Every workplace has hazards, so processing worker have a legal responsibility to look after their workers' safety and protect them against health and safety hazards at work.
- Workplace hazards are any aspect of work that causes health and safety risks and have the potential to harm.
- Some hazards are more likely to be present in some workplaces than others, and depending on work which processing worker will do, there will be hazards that are more or less relevant to his/her work.

Types of workplace hazards

Physical hazards

- It is most common hazards and are present in most workplaces at some time.
- It include: frayed electrical cords, unguarded machinery, exposed moving parts, constant loud noise, vibrations, working from ladders, scaffolding or heights, spills, tripping hazards.

Ergonomic hazards:

- It occur when the type of work seed workers will do by giving strain to their body position and/or working conditions.
- It includes poor lighting, improperly adjusted stacks, frequent lifting, repetitive or awkward movements.

Chemical hazards

- It is when workers are exposed to any chemical preparation (solid, liquid or gas) in workplace.
- It includes cleaning products and solvents, vapours and fumes, carbon monoxide or other gases, gasoline or other flammable materials.

Biological hazards:

- It appears from working with people, animals or infectious plant material.
- It includes blood or other bodily fluids, bacteria and viruses, insect bites, animal and bird droppings.

Managing different work place hazards

Task	Hazards	Plans to eliminate or control the hazards
Restocking shelves and product displays	Products falling down Damaged shelves breaking	Train workers, safety footwear worn by staff Inspect and repair/replace damaged shelves
Frequently lifting and carrying products	Back injuries, overuse injuries of the arms and shoulders	Provide workers with carts, dollies, or hand trucks
Cleaning floors, washrooms, public areas	Working with unknown chemicals Chemicals contacting the skin, eyes	Have safe use information [Material Safety Data Sheet (MSDS)] about each cleaning solution available at the workplace If necessary, have gloves and eyewear available for workers
Restocking storage rooms, moving around the workplace	Slipping and tripping	Remove clutter and waste materials from walking areas Clean up spills that can make the floor slippery for walking
Working alone	Not having anyone to help in case of an emergency	Employer will provide a telephone with which to contact the employer or emergency services
Using electrically powered equipment	Unsafe operation by worker Damaged cord or broken ground pin	Train worker and closely supervise until competent Repair cords, inspect all equipment for damage
Operating forklift truck	Unsafe operation by worker Forklift doesn't function properly	Train worker and closely supervise until competent Maintain the forklift according to the manufacturer's instructions
Collecting seed carts in the parking lot	Being struck by a motor vehicle	Workers must wear high visibility vest
Working around equipment with rotating parts	Long hair and loose clothing getting caught in the rotating parts Fingers, hands or arms getting entangled in the rotating parts	Workers should confine their hair and wear clothing that fits closely to the body Rotating parts should be enclosed by guards provided by the manufacturer
Cutting operations	Cuts Heavy items falling on the feet Foreign objects in the eyes	Workers could wear chain mail gloves, knives could be sharper Workers should wear shoes/boots with protective toe caps Workers may need to wear protective eyewear
Frequently lifting and carrying products	Back injuries, overuse injuries of the arms and shoulders	Provide workers with carts, dollies, or hand trucks

1.5.3 Cleanliness at workplace

- ❖ A clean workplace can be said to as a place where there is all-round protection from germs to health.
- ❖ A clean and organized workplace environment will not just be beneficial for worker but also will help in getting more customers.
- ❖ Cleanliness helps in increasing productivity as dirty place and clean mind cannot work together.
- ❖ Workplaces in many years have realized the importance of clean and healthy surrounding.
- ❖ As workplace is an area where workers spend several hours during their working, so it is advisable that it should remain clean, germs-free and healthy.
- ❖ In workplace, same concept applies as productivity is influenced by environment.
- ❖ A clean environment will increase productivity as people will work with positive mindset that is free of stress.
- ❖ In health, cleanliness is important as hygienic practices contribute to good health of individuals. A dirty environment predisposes people to diseases, which affects their ability to perform activities and it results in less productivity.
- ❖ Seed workers needs to understand the importance of cleanliness in workplace and practice a few workplace hygiene tips so as to maintain a happy and clean office environment.

Signs of poor workplace cleanliness

There are many signs which reflect poor cleanliness at workplace such as:

- ❖ cluttered and poorly arranged work areas
- ❖ untidy or dangerous storage of materials
- ❖ dusty and dirty floors and work surfaces
- ❖ items lying on work paths which are in excess or no longer needed

- blocked or cluttered work area or path
- tools and equipment left in work areas instead of stored properly
- broken pots, containers and damaged materials
- overflowing waste bins and containers
- spills and chemical leaks

Cleanliness and hygiene Rules for seed workers:

- No cluttering of material used

- Always use gloves and proper hand sanitiser
- Make use of rubbish bin and proper damping policy

- Removing of unwanted material from seed room, godowns, fields
- Usage of proper working tools
- Keeping proper stock of chemicals and other materials properly
- Proper spraying in seed rooms for seed storage
- Disposal of boxes and packaging covers immediately

2

Participate in Safety Awareness Campaigns

2.1. Attend fire drills and other safety related workshops organised at the workplace

2.1.1 Introduction

- Fire is a hazard in any part of workplace whose consequences include the threat to lives, health, material and safety of relevant persons, damage to or loss of property and severe interruption to normal working activities or opportunities.
- Managing risk of fire, demands fire safety precautions as per combination of appropriate prevention and protection measures which depends on workarea usage and occupancy, inherent fire risks and legal obligations laid on seed company as seed processing organisation as responsible person.
- Companies have to conduct risk assessments which examines all possible hazards, analyse them carefully and evaluate them as per their likelihood of causing injury and damage and according to possible severity of impact.
- As emergency incidents are quite rare and cannot be predicted, seed workers must be prepared to react without wasting time.
- To minimise the impact, every site worker in such situation must know exactly what to do, for which, detailed emergency management, and repeated exercises or drills are important.

2.1.2 Fire drill

- Fire drills are for more than fires.
- Knowing the evacuation routes of godown/warehouse will help in the event of power outage, gas or chemical leak, or other emergency situation that can occur at workplace.
- Fire drills ensure that workers exit the work area in timely fashion and know their team's designated meeting area, which helps rescue effort in event a processing worker in unaccounted for during an actual emergency at work area.

2.1.3 Organising fire drill

- Committee which includes one representative of every department, where group named one representative from each department as team leader.
- In this, team leader is responsible for ensuring that his/her team evacuates the work area, and he/she must meet team workers at predetermined location outside the workplace.
- Providing evacuation information to all workers before drills begin.
- Telling workers about evacuation path for their work area, and alternative paths in an event, when exit is blocked.
- Creating evacuation signs and post them throughout the workplace.
- Receiving information to workers on how to report a fire and use of fire extinguisher, as well as whom to contact in case of an emergency.
- Developing several fire drill scenarios to assist with evacuation preparedness.
- Giving special attention to process of evacuating of any worker with disabilities or special medical needs.

2.1.4 Managing fire drills

- Conduct regular fire drills, but don't overdo it.
- In case of initial training purposes, hold drills every two weeks for shorter period of time.
- After training period, conduct fire drill after every three months.
- Hold drills more frequently which causes workers to become complacent, where some might refuse to participate.
- Hold surprise fire drill exercise, if workers are too frequent.
- Do not actually block an exit during fire drill or tamper with the workplace.

2.2 Be aware of first aid, evacuation and emergency procedures

2.2.1 First Aid

- First aid is the means to help stabilize a sick or injured person until medical help arrives.
- It can prevent further injury and save lives.
- It is only a first response and cannot take the place of medical help.
- A quick and proper response to poisoning is crucial.

To provide first aid, workers and their family members should know:

- Where to find a list of emergency phone numbers
- Signs and symptoms of poisoning for the pesticides being handled

- The first aid treatment for pesticide poisoning for the pesticides being handled
- Where the pesticide application is taking place and what time the applicator should return
- Where to find a first aid kit

Provide first aid trainings

- While working at site, seed growers and labours involves in certain type injuries.
- It is the duty of seed processing worker to keep/arrange first aid tools at site, so that in case of any emergency, an immediate first aid could be given to the patients using first aid techniques.

First aid kits and attendants

- Make every worker aware of where the first aid kit is located and how to call the first aid attendant if one is required.
- Post signs indicating how to access first aid.
- If a first aid attendant is required, make sure the attendant holds a first aid certificate of the level necessary for that workplace.
- If you require a first aid attendant, train backup attendants. Ensure that enough workers are trained for this responsibility to cover vacations and other absences.

Common site injuries

Common site injuries could be bleeding, burns, eye injury, chemicals inside the eyes wounds and breathing etc.

Control of bleeding

- Direct pressure – use your hand(s).
- Elevate (raise) the limb
- Apply a pad and firm bandage.
- If necessary use clean rags or clothing.

Remember

- Always check circulation below the bandage!
- If there is tingling, numbness or blueness loosen the bandage.

Burns

- Cool the burnt area with cool water for 10-15 minutes.
- If necessary, cover the burn with a clean dressing or plastic wrap before removing person to medical aid.

Remember

- ❖ Do not burst blisters.
- ❖ Do not remove clothing that is stuck.
- ❖ Do not apply creams.

Eye injuries

Foreign bodies in the eye(s)

- ❖ Wash the eye(s) with eyewash or clean water.
- ❖ If the foreign body is stuck to the eye surface, do not attempt to remove it.
- ❖ Place a covering over the eye(s) and send for, or take, the person to medical aid.

Chemicals in eye(s)

- ❖ Wash the eye(s) with clean cool water for at least 15 minutes.
- ❖ Wash from near the nose outwards and always wash under the upper eyelid.
- ❖ Send for, or take, person to medical aid.

Minor wounds

Abrasions, small cuts

- ❖ Clean the wound with mild antiseptic or soap and water.
- ❖ Cover lightly with clean dressing.

Breathing

If a person is breathing but unconscious, turn them onto their side to prevent airway obstruction (tongue or vomit).

Handling/storing of first aid equipment

Immediately and properly examining and treating of injuries save life and is essential to reduce pain and help injured people make a quick recovery. All this is possible at site through first aid kit which should comprise of following:

- ❖ Band-Aids
- ❖ Absorbent compress dressings
- ❖ Sterile gauze pads
- ❖ Adhesive cloth tape
- ❖ Gloves
- ❖ Roller bandages
- ❖ Antiseptic wipes
- ❖ Antibiotic ointment

- Scissors and tweezers
- Sterile saline
- Hydrocortisone ointment

Precautions

- It is the duty of seed worker to check the first aid kit on regular basis, as medications expire, which needs replacement.
- Seed worker should replenish supplies that he/she has used in the past, so make sure that more common treatments like medications, bandages etc are always available.
- If possible use antiseptic solution to clean the scissors and injections.

Reporting first aid procedure at workplace

- Seed worker must provide information about first aid to their superiors so that they should know what to do and whom to contact in case of any emergency.
- Information should be easy to understand, accessible and should take into account the language and literacy levels of the workers.
- Information may be given using verbal methods or visual methods which should include:
  - location of first aid equipment and facilities
  - names and location of persons trained to administer first aid
  - procedures to be followed when first aid is required.

The information and instruction should be provided as part of workers' induction training and when there are any changes, for example in the location of first aid facilities or in the names, locations or contact details of first aiders.

It is the duty of the gardener/seed grower to check that medical person should keep all the details regarding the type of treatment, what he/she is giving along with details of first aid equipment which are dispensed. Medical officers should record maximum information about the patient and type of injury, what the patient is suffering from, with old patient history. This information is important when handing-over the casualty to a health care professional which should cover:

- Name and contact number of casualty;
- Nature of the injury;
- Time and date of incident;
- Details of treatment given

2.2.2 Emergency Procedure

- Emergency procedure is a plan of actions which needs to be done in certain order or manner, in response to specific class of foreseeable emergency, which poses immediate risk to health, life, property, or environment.
- There are many emergencies which require urgent intervention to prevent a worsening of situation, while in some situations, mitigation may not be possible and agencies will offer only palliative care for aftermath.
- An emergency procedure should provide with appropriate and proportionate response to certain situation and should identify responsibilities, actions and resources required to deal with emergency situation.
- Once everything has been drafted, a procedure should be followed by consultative period with those who involved/affected by emergency, and a programme should be set out for testing, training and periodic review.

Workplace emergency

Emergencies at workplace include:

- Floods
- Fires
- Toxic gas releases
- Chemical spills
- Radiological accidents
- Explosions
- Civil disturbances
- Workplace violence

Role of seed processing worker during emergency/evacuation

- Assessing the situation to determine whether an emergency exists requiring activation of the emergency procedures;
- Supervising all efforts in the area, including evacuating personnel;
- Coordinating outside emergency services, such as medical aid and local fire departments, and ensuring their available and notified when necessary;
- Directing the shutdown of workplace operations when required.

Emergency response plans

- Seed workers should be prepared to respond to emergencies such as fires, chemical spills, or natural disasters.
- If an emergency occurs, there will be a need to make quick decisions that will minimize injuries and damage.

- Such decisions are easier if seed grower have already developed an emergency plan.

These are certain guidelines which needs to be followed while developing and implementing an emergency plan which covers:

- List all possible events (for example, serious injuries, fires, explosions, or natural disasters).
- Identify the major consequences associated with each event (for example, casualties, equipment damage, or facility damage).
- Determine the necessary measures to deal with those consequences (for example, first aid, notification of medical authorities, rescue, firefighting, or equipment evacuation).
- Determine what resources will be required (for example, medical supplies or rescue equipment).
- Store emergency equipment where it will be accessible in case of emergency.
- Ensure that workers are trained in emergency procedures and shown where equipment is stored.
- Hold periodic drills at least annually to ensure that employees will be ready to act if an emergency occurs.
- Communicate the plan to everyone involved.

Fire and emergency procedures

It is the duty of seed processing worker to know and understand the following procedures in case of fire:

- Raising the alarm and evacuating people from the work area
- Calling emergency services
- Calling the manager/incharge
- In case of small fire, use fire extinguisher to extinguish the fire

- If the fire is near a fuel tank, do not extinguish the fire, but retreat to safe distance
- Be prepared to direct the fire service to the fire area

2.2.3. Emergency evacuation procedures

- On hearing an evacuation alarm, or on instruction of emergency control personnel, immediately cease all activity and safeguard personal valuables.
- Help any person in immediate danger, but only if required to do so.
- Safeguard any activity/process that results hazardous or suffer damage if left unattended as a consequence of evacuation.
- Act as per directions given by emergency control personnel and evacuate the work area immediately.
- Assist with general evacuation if directed to do so by emergency control personnel.
- Assist with evacuation of disabled occupants.
- In fire, do not try to work in fields.
- Move calmly to required evacuation area and do not leave the evacuation area till everything has been cleared.
- Follow instructions of relevant emergency services personnel.

2.3 Be alert of any events and do not be negligent of any safety procedures to be followed

2.3.1 Following Safety Procedures

Feeling safe is an important aspect of life, so seed workers should clearly follow certain procedures, to keep themselves safe at workplace which includes:

Safety Equipment

- All seed workers should wear the required safety gear while loading or unloading on premises.
- The equipment must also be kept clean and maintained.
- Wearing loose clothing on workplace should be avoided as it gets trapped in machines etc.
- The workers should remain alert while working with moving machinery.

Risk Assessment

- To protect yourself and others, seed workers ahould always be aware of what could go wrong and what you can do to prevent a potentially dangerous situation.

- Never take unnecessary risks and stay calm if you find yourself in perilous predicament.
- Conduct risk assessment to identify and prevent work hazards.

Safety Training

- Safety training is essential when working in godowns.
- Workers must be aware of all safety procedures, especially as they pertain to their specific position.
- Special training should occur periodically to reduce the risks of accidents.

Emergency Evacuation Drills

- Fire and other evacuation drills are excellent way to help everyone understand the best conduct in case of an emergency.
- Workers should be familiar with all emergency exists and know where fire extinguishers are kept and how to use them.
- It's also important to determine which machines must be shut down and which can be left running while evacuating the warehouse.

Working with Tools

- All tools should be kept clean and in perfect working condition.
- Any defects should be immediately reported.
- No tools should be left in a location not meant for storage.

Operating Heavy Machinery

- Machinery should be in full working condition and have all the manufacturer's safety guards installed.
- Anytime a malfunction or defect is noticed, go straight to the maintenance department and review the instruction manual.
- Hazardous manual handling continues to be an underrated risk in many warehouses.
- If the workers are not trained properly in technical aspects of operating certain machinery, in such case avoid such machines.
- Once the work is done, shut down the machine.

Maintaining Cleanliness

- Keeping work area clean does wonders when it comes to preventing accidents.
- A clean and organized area helps to create a better, less stressful work atmosphere.
- Many factors should be taken into consideration, including proper ventilation and lighting fixtures.

2.3.2. Knowing about emergency action plan

An emergency action plan covers required actions which seed processing workers and godown workers must take to ensure safety, in case of fire and other emergencies.

Emergency action plan must include:

- A required method for reporting fires and other emergencies;
- An evacuation policy and procedure;
- Emergency escape procedures and route assignments like floor plans, workplace maps, and safe or refuge areas;
- Details of individuals both within and outside the workplace to contact for additional information/explanation of duties and responsibilities under the emergency plan;
- Procedures for workers who remain to perform or shut down critical workplace operations, operate fire extinguishers or performing other essential services which cannot be shut down for every emergency alarm before evacuating;
- Rescue and medical duties for any workers designated to perform them.

3

Understand Potential Sources of Accidents

3.1 Avoid accidents while using hazardous chemicals, machines, sharp tools and equipments

3.1.1 Chemical hazards

- It is when workers are exposed to any chemical preparation (solid, liquid or gas) in workplace.
- It includes cleaning products and solvents, vapours and fumes, carbon monoxide or other gases, gasoline or other flammable materials.

Since pesticides, herbicides and fumigation requires specialized agrichemicals, so seed processing worker or seed workers should know different hazards and should monitor to see different risk level. Seed workers are exposed to chemicals like pesticides, veterinary drugs, solvents, and oils that causes different long-term hazards.

Commonly used agrichemicals include:

- glyphosate
- aluminium or zinc phosphide – fumigant
- organophosphates
- pyrethroids
- methyl bromide
- sodium fluoroacetate – '1080' baits
- strychnine
- paraquat, diquat
- cresol
- alkaline and acid cleaning agents
- neonicotinoids
- formalin

- nutritional supplements – selenium and copper
- livestock vaccines and drenches

Seed workers need to take care when storing, transporting, using and disposing of chemicals to ensure their own safety; neighbours safety and environment. Any chemical should be treated with extreme caution and should be used as per the instructions as vapours or direct exposure can lead to many acute and chronic health effects.

Accidents due to chemical

The effects of chemical exposure to seed workers while working in warehouse depend on the type of chemical and degree of exposure. If chemicals are swallowed, absorbed through the skin or inhaled as mist, vapour or dust, then there appears a different immediate effect such as:

- poisoning
- headache
- nausea
- vomiting
- diarrhoea
- pinpoint pupils
- dizziness
- fine muscle twitching
- increased bronchial and lacrimal secretions
- skin rashes and irritation
- chemical burns.

3.1.2 Accidents due to vehicles

Warehouse vehicles and implements are involved in many injuries and fatalities, so number of factors that will contribute to risk at warehouse include:

- unrestrained passengers being carried in tray or on trailers
- unsupervised children around moving vehicles
- inexperienced drivers
- poorly maintained vehicles.

3.1.3 Accidents due to sharp tools

Using Sharp Blades

- The most common concern in seed warehouse is when seed workers uses sharp blades or edges which lead to an injury like as cut (laceration, puncture) or amputation.

- Tools or equipment used in warehouse with sharp blades or edges includes:
  - knives and other utensils
  - box cutters, utility knives, safety cutters, etc.
  - scissors
  - equipment with blades or moving parts, such as food processors, mixers, meat grinders, etc.
  - glasses, dishes, etc., especially when broken or chipped
  - food wrap container blades, cans that have been opened, etc.
  - razor blades

Precautions:

To prevent cuts, seed workers should:

- Use right tool for the job.
- Only use the tool only for job it was designed for as knife should not be used as pry bar, can opener, chisel, punch, awl, scraper, or screwdriver.
- Educate and train all workers in safe use of any tool they may use.
- Inspect the tool before use.
- Make sure the blade is sharp as dull blades require more force, increasing the chance of injury.
- Carry one tool at a time, tip and blade pointed down at your side.
- Work in a well lighted space to see what is going on.
- Use cutting board and/or slip-resistant matting to prevent item from sliding on counter.
- Use mechanical device to hold the item.
- Hold the tool with stronger hand.
- Use protective clothing like cut resistant or mesh gloves for holding hand. Wear of safety glasses to protect eyes if blade shatters or breaks.

3.2 Use safety materials such as protective gear, goggles, caps, shoes, etc. (as applicable with workplace)

3.2.1 Introduction

- Seed workers are involved in different types of process right from cutting, harvesting, equipment and vehicle maintenance, loading and unloading of seed bags and placing seed bags on stacks, so for doing such jobs, they should use appropriate cloths, tools and equipment to work.

- As seed workers work in such areas that are not designed for proper working due to various dangers and hazards, so they are advised to use proper clothing and protective equipment while working at workplace.
- Protective Equipment refers to as any specialized equipment or clothing that are used by farmers and farm workers for protection against health and safety hazards, so they are designed to protect many parts of the body such as eyes, head, face, hands, feet, ears, torso.
- The protective equipment not only prevent accidents, but also prevent/ reduce injuries and fatalities.

With respect to equipment and tools to be used at site, farmers or farm workers should ensure that:

- right tool and equipment are available for job
- tools are in appropriate condition
- tools are suitable for working environment
- tool has ergonomic design

Seed workers should see that all clothing and protective equipment must:

- be appropriate for risks involved, without itself leading to any increased risk
- correspond to existing conditions at workplace
- take account of ergonomic requirements and workers state of health
- fit the wearer correctly after any necessary adjustment

Since seed workers are exposed to pesticides and organic dust which serves as most important occupational risks, so to overcome the occupational health hazard, protective clothing/accessories are necessary.

Seed workers should note that protective clothing should be designed in such a way that they are suitable and acceptable when its functional features provides protection to the wearer without causing any health problem or hindrance while working. The designed protective clothing for farm workers reduces occupational health hazards and increases work efficiency.

3.2.2 Selecting required protective clothing or equipment

The type and nature of hazards in workplace will be the main indicator of right type of protective clothing or equipment to be used. The following questions should be considered when assessing whether appropriate protective clothing or equipment is suitable or not:

- Is it appropriate for the risks involved and the conditions at the place where exposure to the risk may occur? In case of eye protection, designed for providing protection against agricultural pesticides will not offer adequate face protection for someone using an angle grinder to cut steel or stone;

- Does it prevent or adequately control the risks involved without increasing the overall level of risk?
- Can it be adjusted to fit the wearer correctly? It is applicable in case of close fitting respiratory protective equipment for which, specialist fit testing is required.
- Has the state of health or needs of those who will be wearing it been taken into account?
- What are the needs of the job and the demands it places on the wearer? For example, the length of time the appropriate protective clothing or equipment needs to be worn/use, the physical effort required to do the job and the requirements for visibility and communication;
- If more than one item of appropriate protective clothing or equipment is being used/worn, are they compatible?

3.2.3 Types of Protective Clothing

Various types of PPE are available for use in the workplace. The Health and Safety Executive provides guidance and general information about types of PPE used in industry, but it doesn't cover specialised and less-used items.

Detailed information should be obtained from suppliers on these more specialised items. Potential users should be involved in the selection of equipment they will be expected to wear and if possible more than one model should be made available to them.

The different types of PPE include:

- Head and scalp protection;
- Respiratory protection;
- Eye protection;
- Hearing protection;
- Hand and arm protection;
- Foot and leg protection;
- Body protection;
- Height and access protection.

Head and scalp protection

There are certain primary purposes of head protection which will help to protect:

- head from fall;

- against falling objects, impact with fixed objects, or wielded weapons;
- head by offering thermal insulation;
- against entanglement and laceration to the head;
- against scalping/entanglement particularly on machinery where injuries are still numerous.
- all forms of head and scalp protection must be suitable, correctly fitted and have an easily adjustable headband, nape and chin strap where appropriate. The relevant standards are BS EN 397 and BS EN 14052.

Eye protection

- The protective equipment for eyes is intended to provide protection against impact, cuts, splashes, mists and sprays.
- All PPE must be regularly cleaned, but this is especially important in the case of eye protection as dirty lenses lead to poor vision and may contribute to accidents.
- Where lenses become scratched, pitted or cracked they should be replaced.
- Users who need to wear corrective lenses (glasses) should have this requirement accommodated in the provision of the PPE to them eg as protective over glasses where appropriate, or in the form of prescription lenses if necessary.
- Where they may be required to wear eye protection on a regular and prolonged basis then any goggles, safety-glasses etc should meet the user's prescription requirements.

Hearing protection

- Assessments carried out under the 'Control of Noise at Work Regulations 2005' will determine whether personal ear protectors are required in the workplace or not, and the noise attenuation required.
- In providing hearing protection, employers should select protectors which are suitable for the working environment and should consider how comfortable and hygienic they are. Like other PPE, hearing protection will need to be compatible with other PPE worn by workers.
- Employers may also wish to provide a range of protectors to allow employees to choose ones which suit them.
- Bearing in mind that the theoretical attenuation is rarely achieved and it is therefore necessary to over-specify the protection.
- When selecting hearing protection, use the detailed noise assessment to determine the attenuation required at High, Medium and Low frequencies and match this against suitable products.

- Bear in mind that where ear plugs are used, training will be needed to ensure that they are used correctly.
- Where ear defenders are used it should be ensured that users do not use music headphones or buds simultaneously.
- For high noise environments, it may be appropriate to specify both plugs and defenders.

Hand and arm protection

Most work requires a degree of manual dexterity and consequently the hands are exposed to a wide range of hazards Risks include cuts, abrasions, heat, cold, chemical contamination, vibration, burns, infection, skin irritation and dermatitis.

Before selecting hand and arm protection, the hierarchy of control measures must be followed. Gloves and gauntlets provide the main form of hand protection against a range of industrial hazards, but other forms of PPE such as mitts, wrist cuffs or armlets may also be used.

In the case of manual handling where there may be a risk of piercing by abrasive, sharp or pointed objects, gloves should be provided where these hazards cannot otherwise be removed, isolated or reduced to an acceptable level. Such gloves are usually made from leather, chain mail, rubber, knitted Kevlar or stout canvas. However, gloves should not normally be worn where there is a risk of them being caught in machinery.

Where chemical exposure is a hazard, and the risk extends to contact with the arms, gauntlets should be specified rather than gloves.

BS EN 14328 is the standard for gloves and armguards protecting against cuts by powered knives while BS EN 407 contains the specifications for gloves intended to protect against thermal risk such as heat and/or fire. BS EN 374 Part 1 covers gloves for protection against chemicals and microorganisms. BS EN 511 covers gloves for protection against the cold. BS EN 388 covers the specification of gloves against mechanical hazards.

Foot and leg protection

- A wide range of safety footwear is available providing protection against many hazards to the feet or legs including crushing, slipping, piercing, temperatures, electricity, chemicals, cutting, and chopping.
- Depending on the hazard various PPE options may be appropriate including safety boots and shoes with protective toe caps and penetration-resistant mid-sole; gaiters; leggings; and spats.

Body protection

- The body protection is required for extended periods of work outdoors to protect against the weather, and to ensure high visibility during work where there is mixed vehicle and pedestrian traffic.

- PPE for body may be required where workers are exposed to extremes of temperature, as well as chemical or metal splash, spray from pressure leaks or spray guns, impact or penetration, contaminated dust, excessive wear, entanglement of own clothing or the risk of drowning.

When choosing body protection the following factors should be considered:

- Thermal comfort, for example, due to sweating;
- Cost and practicality of cleaning;
- Emergency procedures, such as buoyancy or the need to be identified or spotted in hazardous situations;
- Level of hygiene control required;
- Level of personal contamination;
- Personal preference;
- Restriction of movement;
- Storage;
- Temperature and humidity fluctuation;
- Whether the worker is involved in a process that is wet or dry.

Respiratory protection

- It covers equipment that ranges from breathing apparatus and positive pressure powered respirators to protective hoods, close fitting full face respirators, half mask respirators and disposable face masks.
- It is always required to select the correct equipment for risk and individual and to ensure there is adequate training in its use.
- The only form of respiratory protection which is suitable for work in a confined space is breathing apparatus, as other forms of respiratory protection do not provide a source of air or oxygen.
- Face fit testing requirements apply to all close fitting respirators.

Height and access protection

- The range of protective equipment is very wide which covers body harnesses, fall-arrest systems, rescue lifting and lowering harnesses, energy absorbers and lanyards.
- Such PPE is specialised and requires thorough training by competent persons, in user checks as well as correct use.
- Equipment will require periodic inspection by a competent person and anchorage points will normally require periodic testing.

Skin care

- Some occupations are prone to skin disease caused by contact with substances like cutting oils; chemicals, degreasants, glass-fibre and some horticultural agents.
- The main condition is dermatitis, in a variety of forms.
- The use of barrier creams, although traditional in many occupations, should not be considered as personal protective equipment.
- It provides be negligible to protect and is unlikely to be effective in preventing contact dermatitis.
- The benefit of barrier creams is to prevent dirt from becoming so ingrained that it is difficult to remove.
- A properly implemented skin-care hygiene programme, including barrier creams, emollient and skin-conditioning creams can help protect and replace the natural skin oil which is otherwise removed by degreasants and solvents.
- Before undertaking this, advice should be sought from independent skin-care specialists or a dermatologist.

3.3 Handle heavy and hazardous materials with care and using appropriate tools and handling equipment such as trolleys, ladders

3.3.1 Hazardous substances

Hazardous substances are those which, alone or in combination with others, could be or could become:

- toxic
- irritant
- explosive
- flammable
- corrosive
- obnoxious

3.3.2 Possible Hazards of Hazardous substances

While using hazardous substances, possible hazards encountered includes:

- toxicity from absorption, inhalation or ingestion
- spillage
- combustion
- explosion

- corrosion
- extremes of temperature

3.3.3 Impact of Hazardous substances on humans

Hazardous substances can enter in the body as they can be inhaled, ingested or absorbed through the skin where both substances and materials used in technology cause problems.

Hazardous substances when enters the body can cause problems such as:

- Acute illness, which is an immediate reaction to a burn from a substance such as a chemical.
- Chronic illness, which usually develops after exposure over a long period.
- Cancer, which is an example of a chronic illness. Cancer may develop after only a minute
- exposure to a carcinogen. Some substances are classified as known or suspected carcinogens.
- Sensitizing, which is caused by substances that provoke allergies.

3.3.4 Using Hazardous substance in warehouse

Hazardous substances used in seed warehouse should be:

- stored in unbreakable containers
- stored in suitable locked cupboards or other suitable storage to prevent access by unauthorised people
- stored in classifications as per recommended practice
- held in minimum quantities required for specific purposes
- taken into workroom in minimum quantities needed for immediate use
- returned to storage if unused
- opened carefully in case of deterioration or pressurisation of containers

3.3.5 Handling hazardous chemicals

- All workers must be suitably trained in the safe handling of chemicals they are required to use.
- Seed processing worker may need to review the chemicals what are stored in warehouse and to ensure that workers who may come in contact with them are properly instructed.
- Seed processing worker should ensure that site workers are aware of:
    - potential hazards of the substances they use
    - how to control hazards

- how to use personal protective equipment when required
- emergency procedures to follow in the event of a spill, fire or explosion.

3.3.6 Precautions taken by seed workers while working with hazardous substances

Seed workers while working with hazardous substance at ware house or at workplace should follow following rules:

- Workers should be fully instructed in symptoms of chemical ingestion, skin absorption or inhalation so that they recognise danger before the damage is done.
- Safe procedures should be established for the distribution and dispensing of chemicals, whether liquid, solid or in solution.
- Read labels and follow directions on application rates, safety precautions, first-aid directions, and pests or plants to be controlled.
- If there is no label, do not use the chemical.
- Follow application rates and methods of application as directed on labels.
- Use measuring containers to ensure accurate measurements and keep these with the chemical to avoid contamination.
- Ensure equipment used is properly maintained and used in accordance with manufacturer's recommendations.
- Use separate spray equipment for poisonous and nonpoisonous substances.
- Do not reuse the spray unit for other chemicals unless thoroughly rinsed.
- Decide whether weather conditions are suitable for the intended spraying operation.
- Spray in the early morning and when a light breeze is blowing away from areas that could be adversely affected by the spray.
- Ensure that sprays do not drift onto non-target areas.
- Remove any clothes splashed with chemicals.
- Wash all protective clothing thoroughly after use and wash them separately from other clothing.
- Display a prominent and clear warning sign where spraying has been carried out.
- Do not spray when others are present and unprotected, feeling unwell, adequate personal protection is not available, weather conditions are windy or hot, seeds are open, instructions are not clear or not available, preferable methods of control are available

3.3.7 Lifting and Material Handling

- Lifting heavy items is one of the leading causes of injury in workplace where overexertion and cumulative trauma were biggest factors in such injuries.
- Bending, followed by twisting and turning, were more commonly cited movements that caused back injuries.
- Strains and sprains from lifting loads improperly or from carrying loads that are either too large or too heavy are common hazards associated with manually moving materials.
- When workers use smart lifting practices, they are less likely to suffer from back sprains, muscle pulls, wrist injuries, elbow injuries, spinal injuries, and other injuries caused by lifting heavy objects.

Lifting

To lift a load, worker should:

- be close to the load as possible.
- try to keep elbows and arms close to the body.
- keep the back straight during the lift by tightening the stomach muscles, bending at the knees, keeping the load close and centered in front, and looking up and ahead.
- get good handhold and do not twist while lifting.
- not jerk and use smooth motion while lifting.
- not lift too heavy alone, but should take help of someone, to lift the load.

Carrying

For carrying the load, worker should:

- not twist or turn the body, but should move his/her feet to turn.
- face hips, shoulders, toes, and knees in same direction.
- keep load as close to the body as possible with elbows close to sides.
- keep the load down, when feeling fatigued and should take rest for few minutes.

Setting Down

For setting the load, worker should:

- do in the same way as he/she picked it up, but in reverse order.
- bend at knees, not the hips.
- keep his/her head up with tight stomach muscles, and do not twist the body.
- keep load as close to body as possible.
- wait until the load is secure to release handhold.

Index